PROFESSIONAL NEGLIGENCE IN CONSTRUCTION

CONSTRUCTION PRACTICE SERIES

Series editors: Philip Britton and Phillip Capper

FIDIC Contracts
Law and Practice
Ellis Baker, Ben Mellors, Scott Chalmers and Anthony Lavers

Construction Contract Variations
Michael Sergeant and Max Wieliczko
Holman Fenwick Willan LLP

Chern on Dispute Boards
Practice and Procedure
Third Edition
Cyril Chern

Adjudication in Construction Law
Darryl Royce

Delay and Disruption in Construction Contracts
Fifth Edition
Andrew Burr

The Law of Construction Disputes
Second Edition
Cyril Chern

Construction Insurance and UK Construction Contracts
Third Edition
Roger ter Haar QC, Marshall Levine and Anna Laney

Construction Law
Second Edition
Julian Bailey

International Contractual and Statutory Adjudication
Andrew Burr

Remedies in Construction Law
Second Edition
Roger ter Haar QC

For more information about this series, please visit:
www.routledge.com/Construction-Practice-Series/book-series/CPS

PROFESSIONAL NEGLIGENCE IN CONSTRUCTION

BEN PATTEN QC
HUGH SAUNDERS

SECOND EDITION

informa law
from Routledge

Second edition published 2018
by Informa Law from Routledge
2 Park Square, Milton Park, Abingdon, Oxon OX14 4RN

and by Informa Law from Routledge
711 Third Avenue, New York, NY 10017

Informa Law from Routledge is an imprint of the Taylor & Francis Group, an informa business

© 2018 Ben Patten QC and Hugh Saunders

The right of Ben Patten QC and Hugh Saunders to be identified as author of this work has been asserted by them in accordance with sections 77 and 78 of the Copyright, Designs and Patents Act 1988.

All rights reserved. No part of this book may be reprinted or reproduced or utilised in any form or by any electronic, mechanical, or other means, now known or hereafter invented, including photocopying and recording, or in any information storage or retrieval system, without permission in writing from the publishers.

Whilst every effort has been made to ensure that the information contained in this book is correct, neither the author nor Informa Law can accept any responsibility for any errors or omissions or for any consequences arising therefrom.

Trademark notice: Product or corporate names may be trademarks or registered trademarks, and are used only for identification and explanation without intent to infringe.

First edition published 2003 by Spon Press

British Library Cataloguing-in-Publication Data
A catalogue record for this book is available from the British Library

Library of Congress Cataloging-in-Publication Data
Names: Patten, Ben, author. | Saunders, Hugh, author.
Title: Professional negligence in construction / by Ben Patten and Hugh Saunders.
Description: Second edition. | Milton Park, Abingdon : Routledge, 2018. |
   Series: Construction practice series | Includes index.
Identifiers: LCCN 2017056649 | ISBN 9781138553583 (hbk) | ISBN 9781315148649 (ebk)
Subjects: LCSH: Architects—Malpractice—England. | Consulting engineers—
   Malpractice—England. | Construction contracts—England.
Classification: LCC KD2978 .P38 2018 | DDC 343.4107/8624—dc23
LC record available at https://lccn.loc.gov/2017056649

ISBN: 978-1-138-55358-3 hbk
eISBN: 978-1-315-14864-9 ebk

Typeset in Times New Roman
by Apex CoVantage, LLC

Printed and bound by CPI Group (UK) Ltd, Croydon, CR0 4YY

TABLE OF CONTENTS

Table of cases	xi
Table of statutes	xix
Table of statutory instruments	xxi
CHAPTER 1 INTRODUCTION	1
CHAPTER 2 TYPES OF CONSTRUCTION PROFESSIONAL AND THEIR FUNCTIONS	7
Section A: architects	8
Origins	8
Function	8
Professional regulation	9
Architects' contracts	10
Section B: engineers	12
Origins	12
Function	13
Professional regulation	13
Engineers' contracts	14
Section C: quantity surveyors	15
Origins	15
Function	15
Professional regulation	17
Quantity surveyors' contracts	17
Section D: project managers	18
Origins	18
Function	19
Professional regulation	20
Project managers' contracts	20
Section E: building surveyors	21
Origins	21
Function	21
Professional regulation	21
Building surveyors' contracts	22
Section F: other construction professionals	22
Section G: a typical construction project	23

Year 1: preliminary works	23
Year 3: outline scheme design and application for planning permission	23
Year 4: letting the contract	24
Year 5: letters of intent, the building contract, novation	24
Year 5 to year 7: the construction period	25
Year 7 to year 8: defects liability period	26
Year 9 and subsequently: dispute resolution	26
CHAPTER 3 THE OBLIGATIONS OF CONSTRUCTION PROFESSIONALS	27
Section A: the contract	28
The function of the contract	28
When will a contract exist?	28
When will a contract be varied?	30
Section B: express terms	30
Ascertaining the meaning of express terms	30
Common types of problem	31
Express terms relating to services	31
Other express terms	33
Section C: implied terms	35
Section D: the tort of negligence	35
Section E: duty of care – personal injury	36
The general rule	36
Requirement of duty to the client	37
Section F: duty of care – physical damage	38
The general rule	38
No recovery for economic loss	38
No recovery where not just and equitable	39
Examples	40
Section G: duty of care to the client – economic loss	42
The general rule	42
The emergence of concurrent liability	42
Concurrent liability established	43
Section H: duty of care to third parties – economic loss	45
The general rule	45
Exceptions that failed	45
Assumption of responsibility	46
Section I: statutory duties	50
The Defective Premises Act 1972	50
The Construction (Design and Management) Regulations 2015	52
Section J: personal liability	54
Section K: remuneration and other matters	55
Remuneration	55
Disputes over remuneration	56
Plans, documents and copyright	57

TABLE OF CONTENTS

CHAPTER 4	THE STANDARD OF REASONABLE SKILL AND CARE	59
Section A: the *Bolam* test		59
Section B: exceptions to the *Bolam* test		62
Section C: establishing the standard		65
Section D: codes of practice		67
Section E: the band of reasonable actions		68
Section F: outcome, not methodology		69
Section G: seniority, resources and price		71
Section H: delegation		72
Section I: duty to review		74
Section J: skill and care not enough		76
CHAPTER 5	COMMON ISSUES IN A CONSTRUCTION PROJECT	79
Introduction		79
Section A: site investigation and specialist survey		80
Section B: planning and regulation		82
Section C: budgets and estimates		83
Section D: design		87
Introduction		87
Whose obligation?		87
The standard to be achieved		88
Achieving the client's objectives		88
Specific design attributes		90
Coordination and integration of design		92
Novel design		93
The obligation of the reviewer		94
Inspection or supervision by the designer		95
Approval of defective design		95
Section E: the tender process		96
Introduction		96
Specification		97
Bills of quantities		97
Selection of contractors		98
Construction contracts		98
Other contracts		99
Section F: contract administration		100
Introduction		100
Familiarisation		101
Providing information and giving instructions		102
Certification		103
Inspection		106
Reporting to the client		110
Keeping records		110

CHAPTER 6   CAUSATION LOSS AND DAMAGE IN CLAIMS AGAINST CONSTRUCTION PROFESSIONALS   113

Introduction   113
Section A: the essential approach   114
Section B: scope of duty   114
Section C: remoteness   116
Section D: causation   118
    Effective cause   118
    Break in the chain of causation   119
    Outcome dependent upon hypothetical actions   120
Section E: assessment of loss   122
    Costs of rectification or diminution in value?   122
    Other expenditure or costs   124
    Liability to third parties   125
    Personal injury, inconvenience and distress   126
    The construction professional's fees   127
Section F: mitigation   128
Section G: limitation of actions   129
Section H: contributory negligence   132
Section I: contribution   134

CHAPTER 7   INSURANCE   139

Introduction   139
Section A: summary   139
Section B: the contract of insurance   141
Section C: key aspects   144
    The proposal   144
    Notification   146
    Claims and aggregation   148
    Admissions   149
    Disputes   149
Section D: third-party rights   150

CHAPTER 8   EXPERTS, ARBITRATORS AND ADJUDICATORS   153

Introduction   153
Section A: experts in dispute proceedings   154
    Duties and Part 35 of the Civil Procedure Rules   155
    Retainer by instruction   158
    Vulnerabilities   158
Section B: arbitrators   159
    Appointment   160
    Obligations   161
    Vulnerabilities   162
Section C: adjudicators   163
    Appointment   163

Obligations	164
Vulnerabilities	165
Section D: expert determination	165
Appointment	166
Obligations	167
Vulnerabilities	168
CHAPTER 9 DISPUTE RESOLUTION	171
Section A: informal dispute resolution	171
Section B: alternative dispute resolution	172
Section C: adjudication	174
Section D: the pre-action protocol	176
Section E: litigation	181
Section F: arbitration	184
Index	187

TABLE OF CASES

A Jacobs v Morton and Partners (1995) 72 BLR 92 .... 3.89
ABB Ltd v Bam Nuttall Ltd [2013] EWHC 1983 (TCC) 9.30
Abbott v Will Gannon & Smith Ltd [2005] EWCA Civ 198 .... 6.83
Acrecrest Ltd v WS Hattrell & Partners [1983] QB 260 (CA) 5.8, 5.9
ACD (Landscape Architects) Ltd v Overall [2012] EWHC 100 (TCC) 4.28
Adams v Rhymney Valley District Council [2001] PNLR 4 (CA) 4.42
Agouman v Leigh Day (A Firm) [2016] EWHC 1324 (QB) 4.49
Allied Maples Group Ltd v Simmons & Simmons [1995] 1 WLR 1602 (CA) 6.41
Alexander v Mercouris [1979] 1 WLR 1270 (CA) 3.118
Amec Capital Projects Ltd v Whitefriars City Estates Ltd [2005] EWCA Civ 1358 .... 8.55
Andrew Master Hones Ltd v Cruikshank & Fairweather [1980] RPC 16 (Ch) 4.49
Andrews v Schooling [1991] 1 WLR 783 (CA) 3.117
Arab Bank plc v Zurich Insurance Co [1999] 1 Lloyd's Rep 262 (Comm) 7.7
Arnold v Britton [2015] UKSC 36 .... 3.18
Aspect Contracts (Asbestos) Ltd v Higgins Construction Ltd [2015] UKSC 38 .... 9.27
AXA Insurance UK plc v Cunningham Lindsey United Kingdom [2007] EWHC
    3023 (TCC) 6.28, 6.65, 6.74
Bagot v Stevens Scanlan & Co Ltd [1966] 1 QB 197 (QB) 3.77, 3.81
Balfour Beatty Building Ltd v Chestermount Properties Ltd (1993) 62 BLR 12 .... 5.146
Banco de Portugal v Waterlow & Sons Ltd [1932] AC 452 (HL) 6.73
Bank of East Asia Ltd v Tsien Wui Marble Factory Ltd [2001] 1 HKLRD 268 .... 6.83
Bank of Ireland v Faithful & Gould Ltd [2014] EWHC 2217 (TCC) 6.105
Bank of Ireland v Watts Group (UK) plc [2017] EWHC 1667 (TCC) 8.19
Banque Bruxelles Lambert SA v Eagle Star Insurance Co Ltd [1995] QB 375 (CA) 6.25
Barclays Bank plc v Fairclough Building Ltd and ors (1996) 76 BLR 1 (CA) 3.83
Baxall Securities Ltd v Sheard Walshaw Partnership [2002] EWCA Civ 9 .... 3.68
Bella Casa Ltd v Vinestone [2005] EWHC 2807 (TCC) 6.54
Bellefield Computer Services Ltd v E Turner & Sons Ltd [2002]
    EWCA Civ 1823 .... 3.68, 3.84, 5.66
Bernhard Schulte GmbH & Co v Nile Holdings Limited [2004] EWHC 977
    (Comm) 8.71, 8.73
Beumer Group UK Ltd v Vinci Construction UK Ltd [2017] EWHC
    2283 (TCC) 8.55, 8.56
Bevan Investments Ltd v Blackhall and Struthers (No. 2) [1973] 2 NZLR 45 .... 4.31
Biffa Waste Services Ltd v Machinenfabrik Ernst Hese GmbH [2008] PNLR 17 .... 3.66
BL Holdings Ltd v Robert J Wood & Partners (1979) 10 BLR 48; (1980) 12 BLR 1 .... 5.20
Bloomberg LP v Sandberg (a firm) [2015] EWHC 2858 (TCC) 6.102
Blyth v Birmingham Waterworks Co (1856) 11 Ex 781 .... 1.12, 3.46

TABLE OF CASES

Blyth & Blyth Ltd v Carillion Construction Ltd (2001) 79 Con LR 72 ....3.27
Board of Governors of the Hospitals for Sick Children v McLaughlin & Harvey
    plc and ors (1987) 19 Con LR 25 ....4.50, 6.71
Board of Trustees of National Museums and Galleries on Merseyside v AEW
    Architects and Designers Ltd (No. 1) [2013] EWHC 2403 (TCC).............................5.44, 5.76
Bolam v Friern Hospital Management Committee...4.4, 4.7
Bolitho v City and Hackney Health Authority [1998] AC 232 (HL)..4.9
Boorman v Brown (1842) 3 QB 511 (KB) 4.3
Borealis AB v Geogas Trading SA [2010] EWHC 2789 (Comm)..6.33
Brian Warwicker Partnership plc v HOK International Ltd [2005] EWCA Civ 962............6.108
Brickfield Properties Ltd v Newton [1971] 1 WLR 862 (CA) 4.66
British Westinghouse Electric & Manufacturing Co Ltd v Underground Electric
    Railways Co of London Ltd (No. 2) [1912] AC 673 (HL) 6.71
Broster v Galliard Docklands Ltd [2011] EWHC 1722 (TCC) 3.89
Burgess and anor v Lejonvarn [2016] EWHC 40 (TCC); [2017] PNLR 25 (CA) 3.11
Burrell Hayward & Budd v Chris Carnell and David Green (unreported)
    20 February 1992 ....2.32
Cala Homes (South) Ltd v Alfred McAlpine Homes East Ltd (No. 1) [1995]
    FSR 818..3.153
Calvert v William Hill Credit Ltd [2008] EWCA Civ 1427 ....6.9
Camarata Property Inc. v Credit Suisse Securities (Europe) Ltd [2011] EWHC 479
    (Comm) 6.37
Campbell v Edwards [1976] 1 WLR 403 (CA) 8.68
Cannock Chase District Council v Kelly [1978] 1 WLR 1 (CA) 8.42
Caparo Industries plc v Dickman [1990] 2 AC 605 (HL)...3.96
Carillion Construction Ltd v Devonport Royal Dockyard Ltd [2005] EWCA
    Civ 1358...8.54, 9.29
Carillion JM Group Ltd v Phi Group Ltd [2011] EWHC 1379 (TCC)................................6.110
Carillion JM Group Ltd v Phi Group Ltd [2012] EWCA Civ 588 ....6.34
Catlin Estates Ltd v Carter Jonas [2005] EWHC 2315 (TCC) 5.62, 6.44
Cattle v Stockton Waterworks Co (1874–75) LR 10 QB 453 (DC) 3.87
CFW Architects v Cowlin Construction Ltd [2006] EWHC 6 (TCC)....................................4.70
City Inn Ltd v Shepherd Construction Ltd [2010] BLR 473 ....5.146
Clay v AJ Crump & Sons Ltd [1964] 1 QB 533 (CA)..3.51
Clayton v Woodman & Son (Builders) Ltd [1962] 2 QB 533 (CA) 3.51
Cofely Limited v Anthony Bingham and Knowles Limited [2016] EWHC 240
    (Comm) 8.38
Consarc Design Ltd v Hutch Investments Ltd [2002] PNLR 31 ....3.41
Consultant's Group International v John Worman Ltd (1985) 9 Con LR 46.........................4.71
Co-operative Group Ltd v Birse Developments Ltd (in liquidation) [2014] EWHC 530
    (TCC) 6.85
Cooperative Group Ltd v John Allen Associates Ltd [2010] EWHC 2300
    (TCC) 4.7, 4.27, 4.62, 5.12, 6.35, 6.48, 8.18
Copthorne Hotel (Newcastle) Ltd v Arup Associates (No. 1) (1996) 58 Con LR 105 ....4.40, 5.34
Corfield v Grant (1992) 29 Con LR 58..5.152
Costain Ltd v Charles Haswell & Partners Ltd [2009] EWHC 3140 (TCC).........................4.74
D&F Estates Ltd v Church Commissioners for England [1989] AC 177 (HL).....................3.89
Dalgleish v Bromley Corporation [1953] 162 EG 623 ....5.37
Darbishire v Warran [1963] 1 WLR 1067 (CA) 6.71
Department of National Heritage v Steensen Varming Mulcahy (1998) 60 Con LR 33 ......6.25

xii

TABLE OF CASES

Dhamija v Sunningdale Joineries Ltd and ors [2010] EWHC 2396 .... 2.32
Discain Project Services Ltd v Opecprime Developments Ltd (No. 1) [2000] BLR 402 .... 8.55
Donoghue v Stevenson [1931] AC 562 (HL) 3.69, 3.78
Eagle v Redlime Ltd [2011] EWHC 838 (QB) 6.88
Earl's Terrace Properties Ltd v Nilsson Design Ltd [2004] EWHC 136 (TCC) 6.10
East Ham Corp v Bernard Sunley & Sons Ltd [1966] AC 406 (HL) 5.155
Eckersley and ors v Binnie and ors (1988) 18 Con LR 1 (CA) 4.7, 4.11, 5.64, 5.79
Edward Wong Finance Co Ltd v Johnson Stokes & Master [1984] AC 296 (PC) 4.15
Elvanite Full Circle Ltd v AMEC Earth & Environmental (UK) Ltd [2013]
 EWHC 1191 (TCC) 2.62, 5.18
Equitable Debenture Assets Corp Ltd v William Moss Group (1984)
 2 Con LR 1 .... 5.61, 5.72
Esso Petroleum Co. Ltd v Mardon [1976] QB 801 (CA) 3.79
Eurocom Ltd v Siemens plc [2014] EWHC 3170 (TCC) 8.38
Farley v Skinner (No. 2) [2002] 2 AC 732 (HL) 6.64
Fence Gate Ltd v James R Knowles Ltd (2001) 84 Con LR 206 (TCC) 9.23
Fitzroy Robinson Ltd v Mentmore Towers Ltd [2009] EWHC 1552 (TCC) 5.19
Flanagan v Greenbanks Ltd [2013] EWCA Civ 1720 .... 6.32
Forsikringsaktieselskapet Vesta v Butcher [1989] AC 852 (HL) 6.91
Fosse Motor Engineers Ltd v Conde Nast & National Magazine Distributors
 Ltd [2008] EWHC 2037 (TCC) 6.30
Gable House Estates Ltd v The Halpern Partnership (1998) 48 Con LR 1 .... 5.10, 5.34
Gallo Ltd v Bright Grahame Murray [1994] 1 WLR 1360 (CA) 6.25
George Fischer Holding Ltd v Multi Design Consultants Ltd (1998)
 61 Con LR 85 .... 5.61, 5.156
George Hawkins v Chrysler (UK) Ltd and Burne Partnership (1987) 38 BLR 36 .... 2.18
Gilbert & Partners v Knight [1968] 2 All ER 248 (CA) 3.144
Gilbride v Sincock [1955] 166 EG 129 .... 5.53
Greater Nottingham Co-operative Society Ltd v Cementation Piping &
 Foundations Ltd [1989] QB 71 (CA) 3.86
Greaves & Co (Contractors) Ltd v Baynham Meikle & Partners [1975]
 1 WLR 1095 (CA) 4.7, 4.69, 5.48
Gwyther v Gaze (1875), Hudson, Building Contracts (4th ed), Vol. 2, p. 34 .... 2.44
Hadley v Baxendale (1854) 9 Ex. 341 .... 6.18
Hancock v Tucker [1999] Lloyd's Rep PN 814 .... 5.19
Harmer v Cornelius (1858) 5 CB NS 236 (QB) 4.3
Harrison v Shepherd Homes Ltd [2011] EWHC 1811 (TCC) 3.84, 3.118
Harrison v Technical Sign Co Ltd [2013] EWCA Civ 1569 .... 3.52
Hedley Byrne & Co Ltd v Heller & Partners Ltd [1964] AC 465 (HL) 3.80, 3.91
Henderson v Merret Syndicates Ltd (No. 1) [1995] 2 AC 145 (HL) 3.81, 3.90, 3.107
Henry Boot Construction (UK) Ltd v Malmaison Hotel (Manchester)
 Ltd (1999) 70 Con LR 32
Herbosch-Kiere Marine Contractors Ltd v Dover Harbour Board [2012] EWHC 84 (TCC) 9.30
Heskell v Continental Express Ltd and anor [1950] 1 All ER 1033 (KB) 6.28
HIH Casualty & General Insurance Ltd v New Hampshire Insurance Co [2001]
 2 Lloyd's Rep 161 (CA) 7.40
Hi-Lite Electrical Ltd v Wolseley UK Ltd [2011] EWHC 2153 (TCC) 6.32
HLB Kidsons (a firm) v Lloyds Underwriters [2008] EWCA Civ 1206 .... 7.54
Hoadley v Edwards [2001] PNLR 41 (Ch) 6.49
HOK Sport Ltd v Aintree Racecourse Ltd [2002] EWHC 3094 (TCC) 6.10

TABLE OF CASES

Holland Hannen & Cubitts (Northern) Ltd v Welsh Heath Technical Services
  Organisation (1986) 35 BLR 1 (CA)..4.32, 5.49, 5.74
Homepace Ltd v Sita South East Ltd [2008] EWCA Civ 1 ....8.61
Hughes-Holland v BPE Solicitors [2017] UKSC 21 ....6.7
Hunt v Optima (Cambridge) Ltd [2014] EWCA Civ 714...................................3.99, 6.88
IMI plc v Delta Ltd [2016] EWCA Civl 773 ....6.102
Imperial Chemical Industries plc v Merit Merrell Technology Ltd [2017] EWHC 1763 ....8.19
Independent Broadcasting Authority v EMI Electronics Ltd and BICC
  Construction Ltd (1980) 14 BLR 1 ....2.18, 3.99
Invercargill City Council v Hamlin [1996] AC 624 (PC)..6.83
Investors in Industry Commercial Properties Ltd v South Bedfordshire
  DC [1986] QB 1034 (CA)...4.26, 4.56
J Jarvis & Sons Ltd v Castle Wharf Developments Ltd [2001] EWCA Civ 19 ....3.27, 3.111
J Rothschild Assurance plc v Collyear [1998] CLC 1697 (Comm)..................................7.31
J Sainsbury plc v Broadway Malyan [1999] PNLR 286...6.112
John Grimes Partnership Ltd v Gubbins [2013] EWCA Civ 37 ....6.23
John Harris Partnership v Groveworld Ltd [1999] PNLR 697 (TCC)................................5.8
John Laing Construction Ltd v County and District Properties Ltd (1983) 23 BLR 1 ....2.44
Jones v Kaney [2011] UKSC 13 ....8.24
Jones (M) v Jones (RR) [1971] 1 WLR 840 (Ch)...8.66
Jones and anor v Sherwood Computer Services plc [1992] 1 WLR 277 (CA) 8.65, 8.69, 8.75
Kajima UK Engineering Ltd v Underwriter Insurance Co Ltd [2008] EWHC 83 (TCC)..............7.49
Kaliszewska v John Clague & Partners (1984) 5 Con LR 62..5.8
Kensington, Chelsea and Westminster AHA v Wettern Composites [1985] 1 All ER 346...........5.156
Khoo James v Gunapathy [2000] GCA 25 ....4.10
Layher v Lowe [2000] Lloyd's Rep IR 510 (CA)..7.51
Lavarack v Woods of Colchester Ltd [1967] 1 QB 278 (CA) 6.37
Leach v Crossley (1985) 30 BLR 95 (CA)..5.94
Leicestershire CC v Michael Faraday and Partners Ltd [1941] 2 KB 205 (CA).............3.155
Leigh & Sillivan Ltd v Aliakmon Shipping Co Ltd [1986] AC 785 (HL)..........................3.64
Linden Homes South East Ltd v LBH Wembley Ltd [2002] EWHC 536 (TCC)................6.32
Linklaters Business Services v Sir Robert McAlpine Ltd [2010] EWHC 2931 (TCC) 6.85
Livingstone v Rawyards Coal Co (1880) 5 App. Cas. 25 (HL)..6.43
Lloyds Bank plc v McBains Cooper Consulting Ltd [2015] EWHC 2372 (TCC)..............6.98
London Underground Ltd v Kenchington Ford plc and ors (1998) 63 Con LR 1 ....5.60
Maddon v Quirk [1989] 1 WLR 702 (QB)..6.107
Margarine Union GmbH v Cambay Prince Steamship Co Ltd [1969] 1 QB 219 (QB).............3.60
Marks & Spencer plc v BNP Paribas Securities Services Trust Co (Jersey) Ltd
  [2015] UKSC 72 ....3.40
Matrix Securities Ltd v Theodore Goddard [1998] PNLR 290 (Ch)................................4.49
Maynard v West Midlands Regional Health Authority [1984] 1 WLR 634 (HL) 4.8
McConnell v Lynch-Robinson [1957] NI 70 (CA(NI))..6.105
McGhee v National Coal Board [1973] 1 WLR 1 (HL)...6.29
McGlinn v Waltham Contractors Ltd [2007] EWHC 149 (TCC)..............5.154, 5.156, 6.44, 6.46
Merton LBC v Stanley Hugh Leach Ltd (1986) 32 BLR 51 ....5.130
Merton LBC v Lowe (1982) 18 BLR 130 (CA) 4.61
Michael Hyde & Associates Ltd v JD Williams & Co Ltd [2001] PNLR 8 (CA)..............4.11
Michael Salliss & Co Ltd v Calil and William F Newman & Associates (1987)
  13 Con LR 68 ....3.108
Middle Level Commissioners v Atkins Ltd [2012] EWHC 2884 (TCC)............................5.17

TABLE OF CASES

Midland Bank Trust Co. Ltd v Hett, Stubbs & Kemp [1979] Ch 384 ....3.31, 3.79, 4.24, 4.66
Ministry of Defence v Scott Wilson Kirkpatrick and Dean & Dyball
 Construction Ltd [2000] BLR 20 (CA)...5.90, 5.156, 5.157
Moneypenny v Hartland (1824) 1 Car & P 351 ....5.12
Moneypenny v Hartland (1826) 2 Car & P 378 ....5.38
Moresk Cleaners Ltd v Hicks [1966] 2 Lloyd's Rep 338 ....4.61
Mott MacDonald Limited v London & Regional Properties Limited [2007]
 EWHC 1055 (TCC)..8.57
MT Hojgaard A/S v E.ON Climate & Renewables UK Robin Rigg East Ltd
 [2015] EWCA Civ 407...4.73
MT Hojaard A/S v E.ON Climate & Renewables UK Robin Rigg East Ltd [2017]
 UKSC 59..4.73
Multiplex Constructions (UK) Ltd v Cleveland Bridge UK Ltd [2006] EWHC 1341 (TCC).......6.68
Multiplex Constructions (UK) Ltd v Cleveland Bridge UK Limited (No. 6) [2008]
 EWHC 2220 (TCC)..8.4
Munckenbeck & Marshall v Harold [2005] EWHC 356 (TCC)..3.39
Murphy v Brentwood District Council [1991] 1 AC 398 (HL) 3.58, 3.80, 3.89, 3.90
MWB Business Exchange Centres Ltd v Rock Advertising Ltd [2016] EWCA Civ 553 ....3.15
National Justice Compania Naviera SA v Prudential Assurance Co Ltd (The Ikarian
 Reefer) [1993] 2 Lloyd's Rep 68 ....8.19
New Islington and Hackney Housing Association Ltd v Pollard Thomas and
 Edwards Ltd [2001] BLR 74 (TCC) 4.66, 6.83
Neodox Ltd v Borough of Swindon and Pendlebury (1977) 5 BLR 34 ....5.128
Nikko Hotels (UK) Ltd v MEPC plc [1991] 2 EGLR 103 (Ch)..8.67
North Star Shipping Ltd v Sphere Drake Insurance plc [2005] EWHC 665 (Comm) 6.36
Nye Saunders & Partners v Alan E. Bristow (1987) 37 BLR 92 (CA).........1.15, 4.6, 4.17, 5.24, 5.34
Nykredit Mortgage Bank plc v Edward Erdman Group Ltd (No. 2) [1997] 1
 WLR 1627 (HL)..6.80, 6.84
O'Callaghan v Coral Racing Limited [1998] All ER (D) 607 ....9.71
Oldschool and anor v Gleeson (Construction) Ltd and ors (1977) 4 BLR 103 (QB)...................3.52
Ove Arup & Partners International Ltd v Mirant Asia-Pacific Construction
 (Hong Kong) Ltd (No. 2) [2004] EWHC 1750 (TCC) 3.107
Owen Pell Limited v Bindi (London) Limited [2008] EWHC 1420 (TCC)........................8.72, 8.73
Oxford Architects Partnership v Cheltenham Ladies College [2006]
 EWHC 3156 (TCC)...3.39, 6.77
Pacific Associates v Baxter [1990] 1 QB 993 (CA)..3.109
Paice and Springall v MJ Harding Contractors [2015] EWHC 661 (TCC).................................8.55
Pan Atlantic Insurance Co Ltd v Pine Top Insurance Co Ltd [1995] 1 AC 501 (HL)..................7.39
Panamena Europa Navigacion Compania Limitada v Frederick Leyland & Co Ltd
 [1947] AC 428...5.135
Pantelli Associates Ltd v Corporate City Developments Number Two Ltd [2010]
 EWHC 3189 (TCC)..4.28
Parkwood Leisure Ltd v Laing O'Rourke Wales and West Ltd [2013] EWHC 2665 (TCC).........9.22
Payne v John Setchell Ltd [2002] PNLR 7 (TCC) 3.81, 3.89, 3.102
PC Harrington Contractors Ltd v Systech International Ltd [2012] EWCA Civ 1371 ....8.50
Pearson Education Ltd v The Charter Partnership Ltd [2007] 1 BLR 324 (CA)..........................3.74
Perini v Commonwealth of Australia (1980) 12 BLR 82 ....5.146
Perrett v Collins [1998] 2 Lloyd's Rep 255 (CA)...3.53
Pirelli General Cable Works Ltd v Oscar Faber & Partners [1983] 2 AC 1 (HL) 3.80, 6.82
Platform Funding Ltd v Bank of Scotland plc [2009] QB 426 (CA)..4.69

TABLE OF CASES

Pozzolanic Lytag Ltd v Bryan Hobson Associates [1999] BLR 267...................................5.118
Preston v Torfaen Borough Council (1994) 65 BLR 1 (CA)...3.105
Pride Valley Foods Ltd v Hall & Partners (Contract Management) Ltd (No. 1) (2001)
    76 Con LR 1 (CA)..2.47, 5.93, 6.97
Quinn v Burch Bros (Builders) Ltd [1966] 2 QB 370 (CA)...6.25
R v Architects' Registration Tribunal, ex parte Jagger [1945] 2 All ER 131......................2.3
Reeman v Department of Transport [1997] PNLR 618 (CA)..3.97
Rendlesham Estates plc and ors v Barr Ltd [2015] EWHC 3968 (TCC) 3.113
Renwick v Simon and Michael Brooke Architects [2011] EWHC 874 (TCC) 6.89
Richard Roberts Holdings Ltd v Douglas Smith Stimpson Partnership (No. 2)
    (1988) 46 BLR 50 ....2.14, 4.58, 5.67, 6.51
Riva Properties Ltd and ors v Foster and Partners Ltd [2017]
    EWHC 2574 (TCC)..3.14, 5.30, 6.15, 6.26, 6.43, 6.52, 6.69, 8.19
Roberts v Bettany [2001] EWCA Civ 109...6.32
Robinson v PE Jones (Contractors) Ltd [2011] EWCA Civ 9.....................................1.22, 3.81
Royal Brompton Hospital NHS Trust v Hammond and ors (No. 1) [1999] BLR 162 ....5.147
Royal Brompton Hospital NHS Trust v Hammond (No. 3) [2002] UKHL 14..................6.104
Royal Brompton Hospital NHS Trust v Hammond (No. 7) (2001) 76 Con LR 148 ....8.17
Royal Brompton Hospital NHS Trust v Hammond and ors (No. 7) [2001] EWCA
    Civ 206; 76 Con LR 148..5.146
Royal Brompton Hospital NHS Trust v Hammond and ors (No. 9) [2002]
    EWHC 2037 (TCC)..2.48, 4.29, 5.72
Russell v Stone [2017] EWHC 1555 (TCC)..9.56
Ruxley Electronics and Construction Ltd v Forsyth [1996] AC 344 (HL)..............6.47, 6.65
Sahib Foods Ltd (in liquidation) v Paskin Kyriades Sands [2003] EWHC
    142 (TCC)..5.92, 6.97
Saigol v Cranley Mansions Ltd (No. 2) (2000) 72 Con LR 54 (CA) 6.46, 6.48
Saif Ali v Sydney Mitchell & Co. [1980] AC 198 (HL)...4.5
Sainsbury's Supermarkets Ltd v Condek Holdings Ltd and ors [2014] EWHC 2016
    (TCC) 3.107
Samuels v Davis [1943] KB 526 (CA)...4.69
Sansom & anor v Metcalfe Hambleton & Co [1998] PNLR 542 (CA)...................4.26, 8.18
Scheldebouw BV v St James Homes (Grosvenor Dock) Ltd [2006]
    EWHC 89 (TCC)..5.137
Siemens Building Technologies FE Ltd v Supershield Ltd [2009] EWHC 927 (TCC) 6.60
Siemens Building Technologies FE Ltd v Supershield Ltd [2010] EWCA Civ 7 ....6.24, 6.32
Six Continents Retail Ltd v Carford Catering Ltd [2003] EWCA Civ 1790 ....5.156
Skandia Property UK Ltd v Thames Water Utilities Ltd [1999] BLR 338 (CA) 6.73
South Australia Asset Management Corp v York Montague Ltd [1997] AC 191 (HL)....................6.8
Stephenson Blake (Holdings) Ltd v Streets Heaver Ltd [2001] Lloyd's Rep PN 44 ....2.60
St Modwen Developments (Edmonton) Ltd v Tesco Stores Ltd [2006]
    EWHC 3177 (Ch)..3.141
Storey and anor v Charles Church Developments plc (1995) 73 Con LR 1 ....3.81
Stormont Main Working Men's Club and Institute Ltd v J Roscoe Milne
    Partnership (1988) 13 Con LR 127 ....5.54
Sutcliffe v Chippendale & Edmondson (1982) 18 BLR 149..............................5.86, 5.155
Sutcliffe v Thackrah [1974] AC 727 (HL)...4.40, 5.136, 8.75
Taylor v Hall [1870] IRCL 467..2.31
Tesco Stores Ltd v The Norman Hitchcock Partnership Ltd (1997) 56
    Con LR 42...2.13, 3.26

TABLE OF CASES

Thompson v Clive Alexander & Partners (1992) 59 BLR 77 .... 3.118
Thorman v New Hampshire Insurance Co (UK) Ltd [1988] 1 Lloyd's Rep 7 (CA) 7.56
Trustees of Ampleforth Abbey Trust v Turner & Townsend Project Management
    Ltd [2012] EWHC 2137 (TCC) 2.49, 6.41, 6.42
Trustees of London Hospital v TP Bennett & Son (1987) 13 Con LR 22 .... 5.155
Try Build Ltd v Invicta Leisure (Tennis) Ltd (1997) 71 Con LR 140 .... 3.25, 4.58, 5.85
Turner v Garland and Christopher (1853), Hudson, Building Contracts (4th ed),
    Vol. 2, p. 1 .... 5.77
Van Oord UK Ltd and anor v Allseas UK Ltd [2015] EWHC 3074 (TCC) 8.27
Victoria University of Manchester v Hugh Wilson & Lewis Womersley and
    Pochin (Contractors) Ltd (1984) 2 Con LR 43 .... 4.12
Voli v Inglewood Shire Council [1963] ALR 657 .... 4.7
Walter Lilly & Co Ltd v Mackay [2012] EWHC 1773 (TCC) 5.146
Warner v Basildon Development Corporation (1991) 7 Const LJ 146 .... 3.89
Webster v Lord Chancellor [2015] EWCA Civ 742 .... 8.42
Wellesley Partners LLP v Withers LLP [2015] EWCA Civ 1146 .... 3.77, 6.20
West v Ian Finlay and Associates [2014] EWCA Civ 316 .... 6.64
West Country Renovations Ltd v McDowell [2012] EWHC 307 (TCC) 9.61
West Faulkner Associates v Newham LBC (1994) 71 BLR 1 .... 5.124
Whessoe Oil and Gas Ltd v Dale [2012] EWHC 1788 (TCC) 4.28
Whyte and Mackay Ltd v Blyth & Blyth Consulting Engineers Ltd [2013]
    CSOH 54 .... 9.26
William Clark Partnership Ltd v Dock St PCT Ltd [2015] EWHC 2923 (TCC) 5.38, 6.68
Williams v Natural Life Foods Ltd [1998] 1 WLR 830 (HL) 3.93, 3.140
Wimbledon Construction Co 2000 Ltd v Vago [2005] EWHC 1086 (TCC) 9.30

TABLE OF STATUTES

1906
Marine Insurance Act 1906 (c.41) 7.36
1930
Third Parties (Rights Against Insurers)
    Act 1930 (c.25) 7.67
1945
Law Reform (Contributory Negligence)
    Act 1945 (c.28) 6.91
    s.1(1) 6.91
1972
Defective Premises Act 1972 (c.35) 3.113
    s.1(1) 3.113
1977
Unfair Contract Terms Act 1977 (c.50) 3.39
1978
Civil Liability (Contribution)
    Act 1978 (c.47) 6.101
    s.1(1) 6.101
    s.1(3) 6.102
    s.1(4) 6.102
    s.1(6) 6.102
    s.2 .... 6.106
1980
Limitation Act 1980 (c.58) 3.38, 6.77
    s.2 .... 6.77
    s.5 .... 6.77
    s.8 .... 3.38, 6.77
    s.9 .... 3.119
    s.14A 3.77, 6.77, 6.82, 6.83, 6.86, 6.87
1982
Supply of Goods and Services
    Act 1982 (c.29) 3.44
    s.13 .... 3.44, 8.21
    s.14 .... 3.44
    s.15 .... 3.142
1989
Law of Property (Miscellaneous Provisions) Act
    1989 (c.34) 3.38
    s.1 .... 3.38

1996
Arbitration Act 1996 (c.23) 8.29, 8.32
    s.2 .... 8.35
    s.17 .... 8.33
    s.18 .... 8.33
    s.24 .... 8.38, 8.43
    s.29 .... 8.42
    s.33 .... 8.35
    s.42 .... 8.29
    s.43 .... 8.29
    s.44 .... 8.29
    s.52 .... 8.40
Housing Grants, Construction and
    Regeneration Act 1996
    (c.53) 3.146, 9.20
    s.104(1) 9.22
    s.106 .... 9.23
    s.108 .... 9.24
    s.108(2) 8.53, 9.26
    s.108(3) 9.27
    s.108(4) 8.57
1998
Late Payment of Commercial Debts (Interest)
    Act 1998 (c.20) 3.39
1999
Contracts (Rights of Third Parties)
    Act 1999 (c.31) 3.39
2010
Third Parties (Rights Against Insurers)
    Act 2010 (c.10) 7.67
    s.9(2) 7.75
2015
Consumer Rights
    Act 2015 (c.15) 3.39, 3.44, 923
Insurance Act 2015 (c.4) 7.30
    s.3 .... 7.36
    s.7(4) 7.37
    s.11 .... 7.30
    Schedule 1 .... 7.43

TABLE OF STATUTORY INSTRUMENTS

1999
Unfair Terms in Consumer Contracts
    Regulations 1999 (SI 1998/2083) 3.39
2010
Building Regulations 2010 (SI 2010/2214) . 5.21
2015
Construction (Design and Management)
    Regulations 2015
        (SI 2015/51) 3.54, 3.120, 5.65

r.4 .... 3.122
r.5 .... 3.122
r.8 .... 3.123
r.8(1) 3.54
r.9 .... 3.124
r.10 .... 3.124
r.11 .... 3.125
r.12 .... 3.126
r.13 .... 3.126

CHAPTER 1

Introduction

1.1 "Professional negligence" is a mesmeric phrase which has become a term of art in the legal world notwithstanding its imprecision. As commonly understood it refers to the business of suing and defending professional persons, being solicitors, accountants, architects and so forth, for errors or misjudgements made by them in the course of carrying out their work. Lawyers hold themselves out as practising in "professional negligence" and have created bodies and standards to promote and codify their practice. The courts in England and Wales have recognised "professional negligence" disputes as a specific type of dispute, suitable for resolution with the assistance of processes recommended by professional negligence pre-action protocol.

1.2 Yet both "professional" and "negligence" are inexact terms.

1.3 There is no very clear definition of "professional" and the origins of the "professions" are much debated. In antiquity men (and some women) offered their services as doctors, lawyers and architects, but none of them traded as part of a "profession". There was nothing to distinguish the provision of their services from the provision of the services of the blacksmith, the tailor and the cook. The antecedent of profession is the "professio": the act of taking a vow as part of a religious community; the notion of being a member of a "profession" probably has its roots in the experience of joining a religious order. Recognition of professions begins with the decisions of lawyers in the early Middle Ages to band together in Oxford and Paris, ostensibly so as to regulate their members and promote consistent standards, but probably to drive out competition. Doctors soon followed suit.[1] Webster records that the first known use of "profession" is in the thirteenth century.

1.4 The Cambridge Dictionary defines "profession" as "any type or work that needs special training or a particular skill, often one that is respected because it involves a high level of education".

1.5 The characteristics of the professions, as currently understood, are a disciplined group of individuals who adhere to ethical standards. This group positions itself as possessing special knowledge and skills in a widely recognised body of learning derived from research, education and training at a high level, and is recognised by the public as such. A profession is also prepared to apply this knowledge and exercise these skills in the interest of others. A profession arises when any trade or occupation transforms itself through "the development of formal qualifications based upon education, apprenticeship and examinations, the emergence of emergency bodies with powers to admit and discipline members, and some degree of monopoly rights".[2]

1.6 It follows that whether someone is a professional for the purposes of "professional negligence" is a matter of accident and impression.

---

1 See the colourful account in "*The Edge of the World*": Pye [2014] Chapter 6 "Writing the Law".
2 *The New Fontana Dictionary of Modern Thought*: Bullock & Tromley [1999] p. 689.

1.7 It is accidental in the sense that each profession is self-defined and created; it is perfectly open to any group of persons offering skilled services, whether they be fortune tellers or brand consultants, to band together, self-regulate by exclusionary standards and practice requirements and hold themselves out as a profession.

1.8 It is a matter of impression because some degree of public recognition of the group of individuals as a "profession" appears to be a critical threshold; the whole point of creating a profession is to persuade the users of services that these services can only safely be purchased from its members; if the purchasers of the services of fortune tellers and brand consultants blithely ignore the self-regulated group and purchase their services from persons outside it no "profession" is likely to come into being.

1.9 The authors of *Jackson & Powell on Professional Liability* address the issue this way:

> "Professional" is an acquisitive concept, acquisitive of aspirations as well as expectations and liabilities. Yet perceptions as to its contours are indistinct, subjective and constantly changing. The penumbra enlarges with language and context. A professional is not synonymous with a member of a profession and a professional service is professed by others also. The occupations which today are regarded as professions extend far beyond those regarded as such a century ago. They have increased as human knowledge, skill and consequent specialisation have increased. Judicial attempts to define a "profession" recognise that the meaning of the word long ago ceased to be confined to the three learned professions, the church, medicine and the law.[3]

1.10 The authors identify four characteristics or badges of a professional: (1) the service provided is a service of skilled and specialised work which is mental rather than manual and which the professional is able to perform because of a period of specialist theoretical and practical training; (2) a moral commitment to high levels of service and behaviour which is founded in the professional's regard for its profession rather than the professional's narrow economic interest; (3) membership of, or participation in, associations or other collective bodies specific to that profession, which regulate that profession, setting standards and enforcing conduct and (4) status or recognition by the public at large (sometimes augmented by legal privileges) which accords because the professional is a member of a recognised profession and not because of its individual character or abilities.[4]

1.11 "Negligence" is similarly more complicated than it might at first appear.

1.12 In everyday usage "negligence" means carelessness. That plain English meaning is reflected in judicial definition: "negligence is the omission to do something which a reasonable man, guided upon those considerations which ordinarily regulate the conduct of human affairs, would do, or doing something which a prudent and reasonable man would not do".[5]

1.13 However, when lawyers speak of "negligence" they refer not merely to a lack of care, but rather a lack of care in the context where one person owes a legal duty to another to be careful. Indeed, the business of suing and defending professional persons is not primarily concerned with "negligence" at all: it is concerned with *liability*.

1.14 In English law the liability of a professional to its client is largely a matter of contract. The contract between the professional and the client will set out, in varying degrees

---

3 *Jackson & Powell on Professional Liability*, Eighth Edition, paragraph 1–002.
4 The authors of this book paraphrase paragraph 1–003.
5 Per Alderson B in *Blyth v Birmingham Waterworks Co* (1856) 11 Exch. 781 at 784.

of clarity, precisely what it is that the professional is required to do. Whilst it will invariably impose liability upon the professional for what is commonly understood as "negligence", it will generally impose its own definition of negligence and will also generally impose liability for certain non-negligent acts or omissions. It follows that negligence does not exist in the abstract. A professional person is only negligent in respect of the services which its contract requires it to provide.

1.15 Moreover, "negligence", as that word is generally understood by professional negligence lawyers, and as is generally prescribed in the contracts of professional persons, does not mean simple carelessness. It carries a more specialised meaning. "Negligence" is a failure to act within the band of competent actions open to a hypothetical competent member of that profession in those circumstances. This is otherwise known as the *Bolam* test.[6] As will be discussed below, the *Bolam* test is a convenient legal fiction which allows Judges to apply a standard below which conduct will be held to be "negligent". In order to arrive at that standard the court is invariably assisted by experts: knowledgeable members of the profession who not only assist the court in understanding the technical issues but who also provide help with the kinds of judgment which would or would not be taken by a significant body of that profession.

1.16 Whether "professional negligence" provides an entirely satisfactory set of tools for analysing whether the adequacy of the provision of particular skilled services in a construction context is open to question.

1.17 The approach depends for its integrity upon the provider of those services being a member of a profession. In the increasingly complex world of construction, consultants or indeed contractors may from time to time offer the same services as members of the long recognised construction professions (architects, engineers, surveyors) as well as some of the newer and less well established construction professions (project managers, planning consultants). Moreover, as will be seen, the business of construction and the prevalence of construction disputes has created new groups of persons offering "professional" services (experts, claims consultants, arbitrators). How does the court assess the liability of those persons?

1.18 Nor is the practical application of the *Bolam* test without its difficulties. It is all very well to posit the standard of the hypothetical reasonably competent member of the profession, but how is that standard ascertained as a matter of practicality? Is it satisfactory to seek for a uniform minimum standard in a profession which may span a complete range of service level, from the most sophisticated international concern to the one-man-band? How in practice does the Judge (a lawyer) assess whether a particular engineering decision was one which no competent member of that profession would have made? Even if the Judge is assisted by experts, how is their evidence helpful if, as if often the case, the issue is not whether the engineer followed an acceptable approach, but rather the way in which it went about it?

---

6   It is right to bear in mind that, although the term "negligence" has a nasty ring (particularly when it is applied to a professional man), one is not dealing with what might be termed ordinary carelessness. This case concerns a duty which the law says is placed upon somebody who undertakes the responsibility of providing an answer such as Mr Nye had undertaken to do in the circumstances previously outlined/and which are not disputed.
    (per Stephen Brown LJ in *Nye Saunders (A Firm) v Alan E Bristow* (1987) 37 BLR 92 (CA))

1.19 For all these inexactitudes, it must be said that it is difficult to construct a less unsatisfactory set of tools for use in this context. The *Bolam* test is not merely embedded in English law but is adopted in most common law jurisdictions precisely because it is thought to offer the best guide to assessing the liability of professional persons.

1.20 Moreover, in the world of construction projects the language of "professional negligence" now so thoroughly pervades construction contracts that liability is defined in terms of the *Bolam* test. Many if not most contracts between construction professionals and their clients will define the standard of skill and care to be provided by the construction professional by reference to the standard to be expected of a hypothetical competent member of the construction professional's profession.

1.21 The notion of defining the minimum standard of performance expected of a construction professional is habitually borrowed by the draftsmen of design and build contracts (or contracts containing an element of design) where design obligations are framed by reference to the standard of care expected to be observed by a member of a particular profession. Thus, for example, the DOM/2 form of sub-contract provides:

> To the extent that the Sub-Contractor has designed the Sub-Contract Works ... the Sub-Contractor shall have in respect of any defect or insufficiency in such design the like liability to the Contractor, whether under statute or otherwise, as would an architect or, as the case may be, other appropriate professional designer holding himself out as competent to carry on work for such design.[7]

1.22 The use of "the professional" as a special class of persons, whose engagement creates a particular set of legal obligations has even found its way into the wider law. As Jackson LJ observed in *Robinson v PE Jones (Contractors) Ltd*:[8]

> Henderson's case is now taken as the leading authority on concurrent liability in professional negligence...
> In my view, the conceptual basis upon which the concurrent liability of professional persons in tort to their clients now rests is assumption of responsibility. It is perhaps understandable that professional persons are taken to assume responsibility for economic loss to their clients. Typically, they give advice, prepare reports, draw up accounts, produce plans and so forth. They expect their clients and possibly others to act in reliance upon their work product, often with financial or economic consequences.
> When one moves beyond the realm of professional retainers, it by no means follows that every contracting party assumes responsibilities (in the Hedley Byrne sense) to the other parties co-extensive with the contractual obligations...[9]

1.23 "Professional negligence" is therefore here to stay. Our task is to explain how it works.

1.24 In that regard, the title of this book is doubly misleading. Although there is extensive discussion of what is meant by "negligence" and the circumstances in which it will and will not be found to have occurred, the book is really concerned with *liability*: when and in what circumstances will a "construction professional" be liable to its client? Secondly, whilst the book considers the liabilities of a very wide spectrum of persons who provide construction services, it is not concerned, or at least not primarily concerned, with the

---

7 Clause 5.3.1.
8 [2011] EWCA Civ 9, 2012 QB 44. See the discussion in Chapter 3.
9 Paragraphs 74 to 76.

difficult question of whether these persons are members of a "profession". Instead it seeks to focus on the way in which the courts are likely to approach the liabilities of persons offering particular skills or services in the business of construction.

1.25 In Chapter 2, we discuss the differing types of construction professional and their role in construction projects. Whilst some types of construction professional have been recognised for millennia, others are new and reflect the way in which changing technology and economics have brought both complexity and new standards of competence to construction projects.

1.26 In Chapter 3, we discuss the obligations of construction professionals. Whilst reference has been made above the primacy of contractual obligations as being the touchstone for liability, other legal obligations arise between construction professionals and their clients and construction professionals and other persons. Mostly these obligations arise in the tort of negligence but sometimes they are the creature of statute. Just as construction projects have become more complex technically and economically, so they are more regulated and the modern construction professional operates in a web of obligations rather than in pure, bilateral relationship.

1.27 Chapter 4, is concerned with the standard of care which the law posits that a construction professional will owe to its client and (in an appropriate case) to other persons. As indicated the standard of care and the way in which it can be demonstrated (or shown to have been unmet) is a legal construct governed by legal rules and guidelines all of which is anchored in an appreciation of what the relevant profession itself expects of its members.

1.28 Chapter 5, which is the longest chapter in the book, is concerned with examples of where and what circumstances situations may arise where a construction professional becomes liable to its client or occasionally another person. This chapter attempts to follow the broad chronological outline of a typical project, picking up instances where such liability sometimes occurs. By necessity it speaks to construction professionals generally, rather than professionals from a particular discipline, concentrating on the tasks which are to be performed at any particular time.

1.29 In Chapter 6, we examine an often neglected aspect of professional negligence: where a construction professional is liable to its client or a third party what, exactly, is it liable for? How is the financial value of that liability ascertained? Where someone alleges that a construction professional has caused loss and damage what has to be proven, by whom and in what way? A construction professional may be in breach of its contract with its client, but not have caused that client loss. By contrast the smallest and most innocuous departure from its obligations may have caused enormous losses. How does English law navigate these waters?

1.30 Professional indemnity insurance is critical to enabling construction professional to offer its services to the client and to giving the client the confidence to engage the professional in the first place. Chapter 7 considers how professional indemnity insurance works, what obligations are undertaken by the insurer and the insured and how disputes between them are resolved.

1.31 In Chapter 8, we consider the functions of construction professionals who are not engaged to carry out work on a project, but rather who are engaged to provide their opinion or adjudication on matters arising out of a project. Construction professionals increasingly act as experts, adjudicators and arbitrators. What are the obligations of the construction professionals which take on these roles? Can they be "negligent" and if so how? Can they be sued for something other than negligence, and if so in what circumstances and for what sums?

1.32 Lastly, Chapter 9 is concerned with the dispute resolution process. Some construction professionals will never have a claim made against them, but all construction professionals are involved, at one time, in disputes with their clients. How are such disputes resolved without the need to engage lawyers? Where lawyers are engaged how can disputes be resolved without the need to go to court? Where a case against a construction professional goes to court, who does what and how long will it take?

CHAPTER 2

Types of construction professional and their functions

Section A: architects	8
Origins	8
Function	8
Professional regulation	9
Architects' contracts	10
Section B: engineers	12
Origins	12
Function	13
Professional regulation	13
Engineers' contracts	14
Section C: quantity surveyors	15
Origins	15
Function	15
Professional regulation	17
Quantity surveyors' contracts	17
Section D: project managers	18
Origins	18
Function	19
Professional regulation	20
Project managers' contracts	20
Section E: building surveyors	21
Origins	21
Function	21
Professional regulation	21
Building surveyors' contracts	22
Section F: other construction professionals	22
Section G: a typical construction project	23
Year 1: preliminary works	23
Year 3: outline scheme design and application for planning permission	23
Year 4: letting the contract	24
Year 5: letters of intent, the building contract, novation	24
Year 5 to year 7: the construction period	25
Year 7 to year 8: defects liability period	26
Year 9 and subsequently: dispute resolution	26

SECTION A: ARCHITECTS

Origins

2.1 The term "architect" derives from the Latin "architectus", which in turn derives from the Greek "arkhi" and "tekton": chief builder. For the ancients (and indeed until the Renaissance) titles such as "architect" and "engineer" were used interchangeably to apply to master builders, often artisans such as masons and carpenters, who had acquired seniority and experience in substantial construction projects such as castles and churches.[1] Design skills were learned in apprenticeship and through experience. Arguably it was the increasing availability of cheap paper which permitted the production and repetition of construction drawings which allowed the emergence of modern architecture as an academic discipline.[2]

2.2 The emergence of a distinct class of person known as "the architect" and thus the architect's profession is a relatively modern phenomenon. In the United Kingdom that emergence is generally traced back to Sir John Soane who insisted that the architect separate himself from the day to day business of carrying out the building works opining, "with what propriety can his situation and that of the builder, or the contractor, be united?"[3] In 1834 a group of prominent architects formed the Institute of British Architects in London, which became known as the Royal Institute of British Architects in London after the grant of its royal charter in 1837. The Charter set out the purpose of the Royal Institute to be: "… the general advancement of Civil Architecture, and for promoting and facilitating the acquirement of the knowledge of the various arts and sciences connected therewith…". In 1892, it became the Royal Institute of British Architects, or RIBA.[4]

Function

2.3 The Oxford English Dictionary definition of "architect" is "a person who designs buildings and also in many cases supervises their construction". However, the English courts adopted a longer functional definition:

> An "architect" is a person who possesses, with due regard to aesthetic as well as practical considerations, adequate skill and knowledge to enable that person: (i) to originate, (ii) to design and plan, and (iii) to arrange for and supervise the erection of such buildings or other works calling for skill and design in planning as they might, in the course of their business, reasonably be asked to carry out or in respect of which they offer their services as a specialist.[5]

2.4 As construction projects became more intricate the scope of the architect's function expanded. In more recent times the functions of the architect were stated in *Hudson*[6] to be as follows:

(i) To advise and consult with the employer (not as a lawyer) as to any limitation which may exist as to the use of the land to be built on, either (inter alia) by

---

1 *The Culture of Building*, Davis, [2006].
2 *Medieval Architectural Drawing*, Pacey, [2007].
3 The Architect, Kostof, [1977].
4 RIBA Guide to its Archive and History [1986].
5 *R v Architects' Registration Tribunal, ex p. Jaggar* [1945] 2 All ER 131 at 134, citing the criteria employed by the statutory tribunal created by the Architects Registration Act 1938.
6 *Hudson's Building and Engineering Contracts*, 11th Edition. A revised formulation of this statement has been included in the 13th Edition to reflect the services of both Architects and Civil Engineers.

restrictive covenants or by the rights of adjoining owners or the public over the land, or by statutes and by-laws affecting the works to be executed.
(ii) To examine the site, sub-soil and surroundings.
(iii) To consult with and advise the employer as to the proposed work.
(iv) To prepare sketch plans and a specification having regard to all the conditions which exist and to submit them to the employer for approval with an estimate of the probable cost, if requested.
(v) To elaborate and. If necessary, modify or amend the sketch plans as he may be instructed and prepare working drawings and a specification or specifications.
(vi) To consult with and advise the employer as to obtaining tenders, whether by invitation or by advertisement, and as to the necessity or otherwise of employing a quantity surveyor.
(vii) To supply the builder with copies of the contract drawings and specification, supply such further drawings and give such instructions as may be necessary, supervise the work, and see that the contractor performs the contract, and advise the employer if he commits any serious breach thereof.
(viii) To perform his duties to his employer as defined by any contract with his employer or by the contract with the builder, and generally to act as the employer's agent in all matters connected with the work and the contract, except where otherwise prescribed by the contract with the builder, as, for instance, in cases where he has under the contract to act as arbitrator or quasi-arbitrator.

2.5 As will be seen, this list is both inadequate to describe the many functions which an architect may be called upon to perform, but also too prescriptive. Most modern contracts between architects and their clients will contain detailed accounts of the services which the architect is required to provide and whilst the *Hudson's* statement may serve to provide general headings for the groups of functions at different stages of a construction project, it does not do justice to the breadth of the role and the sophistication of the services. Notably absent from *Hudson's* list, for example, is the architect's usual role in substantial projects as design team leader, responsible for coordinating design generally, including the design of other consultants.

2.6 The list is too prescriptive because it assumes that an architect will be engaged on a traditional contract. In respect of substantial projects modern building practice has moved away from the traditional project where the architect was the dominant construction professional guiding the client throughout the life of the project from its earliest conception to the correction of the last defect. In large modern projects the architect may be merely one part of the design team, supervised by a project manager who has taken on the architect's traditional role as the client's chief adviser and agent. Even the scope of the architect's design functions may be limited, the architect being called in to design to a particular stage or even in respect of particular parts of the project.

Professional regulation

2.7 In the United Kingdom, in order to trade as an "architect" a person must be registered by the Architects' Registration Board[7] ("ARB") which requires the applicant to have

---

7 Section 20 of the Architects Act 1997.

to have passed a recognised examination and to have certain practical experience. "Trading"[8] means holding out architectural services as a material part of a business and trading as an architect when not registered is a criminal offence. Usually practising architects also belong to the RIBA which is the most important of the architects' professional bodies.[9] Both the ARB and the RIBA have prescribed codes of conduct[10] and the RIBA produces the most important form of contract, the RIBA Standard Form of Agreement.

2.8 There is no legal requirement for architects to contract in any particular form and the normal laws of contract will apply to any agreement between an architect and his client. Thus although it is prudent for such agreements to be in writing, or at least evidenced by an exchange of correspondence, agreements are often oral or partially in writing and partially oral. Similarly, although it is prudent for all terms of a contract to be agreed before the commencement of services it is common practice for the substantial terms to be agreed at the outset and for remaining matters to be implied according to custom and practice or to be agreed subsequently. This is notwithstanding the requirements in paragraph 2.7 of the RIBA Code of Conduct that a member undertake:

> when making an engagement, whether by an agreement for professional services, by a contract of employment or by a contract for the supply of services and goods, to state whether or not professional indemnity insurance is held and to have defined beyond reasonable doubt and recorded the terms of the engagement including the scope of the service, the allocation of responsibilities and any limitation of liability, the method of calculation of remuneration and the provision for termination and adjudication.

2.9 The RIBA publishes a number of versions of "Standard Form of Agreement for the Appointment of an Architect"[11] which is particularly used in respect of small to medium sized projects and which is often cannibalised to form the basis of appointments on larger projects.[12] It also provides educational services in the form of guidance in respect of various aspects of architectural practice and support services such as advice on matters of practice and insurance.

Architects' contracts

2.10 The architect's contract with its client is the starting point for an understanding of the architect's obligations. Each contract will of course vary depending on the circumstances, but they tend to follow a similar pattern.

2.11 The agreement will set out the identities of the parties, a description of the project, the precise nature of the services to be provided by the architect, the fees and expenses to be paid (usually on a work stage basis), any additional services, and lengthy conditions of engagement under which the architect may be required to do certain things (such as

---

8 The Act covers architects employed by others but does not cover those who operate occasionally or as a hobby. Bodies corporate, firms or partnerships are covered in the same way as individuals.

9 Note should also be made of the Association of Consultant Architects which is made up of consultant architects in private practice.

10 It should also be noted that both bodies exercise disciplinary functions over their members whilst the RIBA runs a conciliation scheme for those dissatisfied with the services of a member architect.

11 See "A Guide to the RIBA Forms of Appointment 1999" (RIBA Publications 1999) which explains the differences between the different forms. There are a number of variations published under RIBA Agreements 2010.

12 Another popular standard form is ACA SFA 2012, published by the Association of Consultant Architects. This is an update of SFA/99.

maintain professional liability insurance) and where the rights of the parties in particular situations may be set out (for example, the right to terminate).

2.12 As set out above, different RIBA standard forms remain the contract of choice for small to medium sized projects. However, in relation to substantial projects, and in relation to design and build contracts, it is common for bespoke contracts to be employed. These may borrow elements from standard form contracts (commonly the RIBA work stages) but also impose obligations and responsibilities which are generally more advantageous to clients.

2.13 It is particularly important for architects accurately to define the extent of their services. On a traditional contract this may be straightforward. The RIBA classification of stages is now broadly adopted. This sets out in detail the functions which an architect is expected to perform at different points in the life of a building project. However, for a more complicated or unusual contract there is often a risk that failure accurately to define roles and responsibilities will lead to future disputes.[13]

2.14 However detailed the contract as to the services which an architect is required to perform there will always be areas of uncertainty. Whilst a description of the architect's responsibilities by reference to different work stages reduces the room for error, it cannot provide a comprehensive account of the architect's responsibility. Thus, the contract may set out that the architect is required to undertake design at a specific point and to coordinate its design with the design of others, but precisely what counts as the architect's design and what design is the responsibility of others can often only be ascertained by a careful analysis of the type of work and its relationship to the architect's contract. In *Richard Roberts Holdings Ltd v Douglas Smith Stimson Partnership (No. 2)*[14] the defendant architects believed that they had discharged their duty to the employer by introducing him to a contractor supplying tank linings. When the linings failed they declined to accept responsibility. However, the court held that the architect's responsibility extended to the whole of the works and thus included the supervision of the lining works.

2.15 An architect's written contract will invariably contain an express term that the Architect exercise "reasonable skill and care".[15] In the absence of an express term – for example because the contract is oral or is only partially recorded in writing – an obligation to carry out the architects' services with reasonable skill and care will be implied by law.[16]

2.16 It is open to the parties to attempt to agree an obligation for some higher degree of skill and care. This is sometimes seen in the use of phrases such as "utmost skill and care" and "highest degree of skill and care". However, in practice such phrases do not tend to add to the architect's obligations – it is too difficult to decide what these phases mean. In practical terms, unless an architect agrees to guarantee that his design will work in a certain way or the building will be constructed to a certain cost (see below) the benchmark by which its

---

13 One of the most frequent disputes is whether an architect who has undertaken the responsibility to design works has also undertaken the responsibility to supervise their construction. This was one of the key issues in *Tesco Stores Ltd v Norman Hitchcock Partnership Ltd* [1997] 56 Con LR 42.

14 (1988) 46 BLR 50.

15 In clause 2(1) of the RIBA SFA/99 the obligation is expressed as "The Architect shall in performing the services and discharging the obligations under [this Agreement] exercise reasonable skill and care in conformity with the normal standards of the architects' profession".

16 Section 13 of the Supply of Goods and Services Act 1982.

performance is to be judged is the exercise of such reasonable skill and care as would be exercised by a reasonably competent member of its profession.

2.17 That said, it is open to an architect not merely to agree to exercise reasonable skill and care but to guarantee that his design will perform to a particular standard. Such warranties are comparatively rare because of the very onerous obligations they place upon the architect: the architect is guaranteeing that, irrespective of circumstances beyond its control, its design will perform to a particular standard. For this reason, English courts have been reluctant to construe the contracts of architects (or other professional persons) as warranting absolute results.

2.18 There are rare instances where the courts have decided that on the facts of a particular case (and irrespective of the architect's duty of reasonable skill and care) the architect provided a warranty that its design would be "reasonably fit for the purpose for which it is intended".[17] Special facts are needed for such an implied warranty to arise. They will generally be confined to circumstances where the court finds that, because of the particular importance of way in which the design was intended to function, the importance which the employer attached to its success and the claims that the architect made on its behalf, the parties intended that such an obligation should exist.[18]

SECTION B: ENGINEERS

Origins

2.19 As with architects, the origin of the term engineer is that was term of art applied generally to a master builder. It derives from the Latin "ingeniator", which in turn combines "ingeiare" (to contrive or devise) and "ingenium" (cleverness). Unlike architects, the use of the term engineers, at least in the United Kingdom, was not confined to the construction design. It acquired very broad currency as referring to any skilled person with an ability to address mechanical or other technical problems.

2.20 "Engineer" in English usage can be traced back to as early as 1325 when "engine'er" was used to describe a soldier skilled in the operation of catapults and other military machinery.[19] With increased use of machinery in the eighteenth century the term became more narrowly applied to fields in which mathematics and science were applied to these ends. Similarly, in addition to military and civil engineering, the fields then known as the mechanic arts became incorporated into engineering.

2.21 Whilst the Victorian notion of the engineer as the master builder has survived in the form of describing the contract administrator in some suites of contracts as "the Engineer", the modern role of the engineer is very much that of a designer and/or consultant, responsible for designing and/or advising on some aspects of the construction, but generally acting as a member of the design team under the overall direction of the architect, contract administrator or project manager.[20]

---

17 See *IBA v EMI Electronics Ltd and BICC Construction Ltd* (1980) 14 BLR 1.
18 See *George Hawkins v Chrysler (UK) Ltd and Burne Associates* (1987) 38 BLR 36.
19 Oxford English Dictionary.
20 Certain standard form contracts intended for specialist engineering or process plant works use the term "Engineer" more literally because engineering expertise is necessary for administering the contract – see paragraph 2.25 below.

Function

2.22 In modern usage the term "engineer" can be applied to so many different professionals in so many different situations that it is incapable of precise definition. Commonly engineers in the construction industry have been defined as non-architects carrying out professional duties analogous to those of architects.[21]

2.23 The most important engineering function is design. What distinguishes engineering design from architectural design is that that the former is generally concerned with the structural, mechanical or chemical attributes of the structure which is being designed. These are matters largely outside the architect's sphere of expertise and the architect will look to the engineer for advice in respect of them. In most types of building project, the design of a structure will thus be a collaboration between the architect and one or more engineers. It is common for such projects to be described in terms of the architectural and the structural design, or the architectural design and the services design.

2.24 Under the general heading of "engineer" lie the specialisms. The two most common and most important for construction projects are civil and structural engineers, who provide expertise as to the structural requirements and integrity of buildings and materials, and mechanical and electrical engineers, who advise as to the provision and effecting of water, air, heating and electrical services to a building. However, just as the designer of a modern building must take account of a multitude of highly technical considerations relating to the building's environment, requirements and fabric, so there are a multitude of consultants offering engineering services in areas such a soil mechanics, concrete and stone behaviour, cladding, paint, fire protection and noise attenuation (to name but a few).

2.25 This is not to say that engineers do not carry out the entire design for some structures. Projects where the structural properties are central to the purpose of the project (bridges, roads, tunnels and so forth) may all be designed by engineers without the involvement of an architect. Some of the standard forms of contract designed for large construction projects of this nature consequently put the engineer at the heart of the design and administration of the project. Moreover, there is no reason why an engineer should not act in the role of project manager or contract administrator, particularly if the subject matter of the project lends itself to the application of engineering judgment.

Professional regulation

2.26 Whilst not as tightly regulated as architects, there are a number of important professional bodies to which engineers belong as members and which are run for the benefit of engineers. In particular, the Institution of Civil Engineers ("ICE") has traditionally operated in a similar manner to the RIBA as the flagship institutional body representing engineers. Whilst not the beneficiary of a royal charter, membership of the ICE enables an engineer to describe himself or herself as a Chartered Civil Engineer. The ICE operates a code of professional standards and by-laws and is responsible for publishing one

---

21 See the definitions in *Jackson & Powell* (Eighth Edition) paragraph 9–003 and *Keating* (10th Edition) paragraph 14–145. The 12th Edition of *Hudson* cites a nineteenth-century definition of Civil Engineer as "one who professes knowledge of the design and construction of works such as bridges, docks, harbours, canals, railways, roads, embankments, water, drainage and gas works and factories".

of the major forms of civil engineering contract, the ICE Conditions. The Association of Consulting Engineers ("ACE") is the United Kingdom's leading trade association for engineering, technical and management consultancies with approximately 650 member firms. It produces the ACE suite of contracts. The Engineering Council is the national registration authority for professional engineers.[22]

Engineers' contracts

2.27 As with architects, there is no legal requirement for a particular form of contract to be entered into by an engineer, although it is common for various types of standard form to be employed. The most commonly employed are the ACE Conditions of Engagement which come in a number of variations which illustrate the breadth of responsibilities commonly accorded to engineers. One of the best known is the ACE Conditions of Engagement Agreement A(1) [2002] for use where the consulting engineer is engaged as lead consultant.[23]

2.28 Because the services provided by engineers are so varied, depending on their discipline, it is less usual (although not uncommon) for the contracts of engineers to be in bespoke form. The usual situation is for the parties to adopt one of the standard form contracts and then adapt it. Often distinguishing between civil and structural engineers, mechanical/electrical engineers and other engineers who fall under the general rubric of consultant, these cover a series of arrangements where the engineer is (i) lead consultant, (ii) not lead consultant but directly engaged by the client, (iii) where the engineer provides design services of a design and construct contractor, (iv) where the engineer provides reporting and advisory services, (v) where the engineer is a project manager and (vi) where the engineer has a particular role under the Construction (Design and Management) Regulations. Additional variations cover minor works and sub-consultancy cases.

2.29 Generally speaking, these standard form contracts will follow the same structure. They commence with a memorandum of understanding which identifies the parties. This is followed by the conditions of engagement themselves and after these there are set out the services which the engineer provides. Although the engineer's responsibilities will generally be clear, as with architects it is important accurately to define these obligations. Unlike architects, there is no equivalent of the RIBA work stages and the scope of the responsibilities will be tailored to the particular functions which the engineer has agreed to provide. In respect of some engagements, for example the provision of structural advice in a building project, the engineer's responsibilities will be couched in terms of providing all reasonably required civil and structural design for particular work stages, together with ancillary obligations. For mechanical and electrical services similar arrangements may be set out in the contract. By contrast the contract for specialist design work in respect of some particular aspect of construction works may be limited by reference to the particular aspect of the works (for example, piling) upon which advice is sought.

---

22 Other important professional bodies include the Institution of Structural Engineers ("I.Struct.E"), the Institution of Mechanical Engineers ("I.Mech.E") and the Institution of Electrical Engineers ("IEE").

23 It should be noted that this long running series has recently been replaced by the 2017 ACE suite of agreements. The ICC suite of agreements is also popular, particularly for consultancy work. IET produces its own suite of agreements, with the IET Model Form of General Conditions of Contract being the most widely used.

2.30 Although it is open to an engineer to warrant that works of his design or constructed under his supervision will perform to a certain standard, such obligations are rare and the usual professional negligence dispute concerns the engineer's performance of his obligation to use reasonable skill and care.[24]

SECTION C: QUANTITY SURVEYORS

Origins

2.31 Quantity surveyors first emerged as a recognised discipline in the early Victorian era when the estimation of cost became so specialised that it could no longer be performed with sufficient accuracy by architects and engineers. The nineteenth-century case of *Taylor v Hall*[25] described the functions of a quantity surveyor as "taking out in detail the measurements and quantities from plans prepared by an architect for the purpose of enabling builders to calculate the estimates for which they would execute the plans".

Function

2.32 The traditional function of the quantity surveyor is narrow compared with that of architects and engineers. Whereas the architect or engineer is generally required to take decisions as to design, instructions to contractors or certification, the traditional role of the quantity surveyor is generally confined to advising on matters of cost:

> All that a quantity surveyor can do is (a) to check that tenders of contractors or sub-contractors are reasonably priced before he recommends acceptance; (b) to measure work executed accurately; (c) to exercise vigilance in the valuation of variations or the checking of the valuation of variations submitted to him by the contractor; (d) make a fair assessment of any additional sums which may be due to a contractor as a result of extension of time, acceleration instructions and so forth. Moreover in assessing the cost of work done it is generally no part of the obligation of a quantity surveyor to assess its quality. He is not charged with identifying defects or other contractual non-compliance.[26]

2.33 It is this narrowness of function (and thus the limited consequences of a quantity surveyor carrying out his functions negligently) which partly explains why there have been relatively few professional negligence actions taken against quantity surveyors in comparison with architects and engineers. Moreover, as will be explained, even where the consequences of a quantity surveyor's error are potentially significant (for example a gross over valuation of work carried out), the employer will often have a remedy against the contractor under the terms of the contract which will mean that he suffers no loss.

2.34 As the management of building costs has become both more complex and more sophisticated the quantity surveyor's role has expanded beyond the calculation of quantities.

---

24 Again this duty will be implied by law if not expressly contained within the contract. In the ACE Conditions of Engagement it is contained at paragraph 2.4.
25 [1870] 4 I R C L 467 at 476.
26 per Mr Recorder Jackson QC in *Burrell Hayward & Budd v Chris Carnell and David Green* [unreported] 20 February 1992. A quantity surveyor will also not normally owe a duty to identify defective work – see *Dhamija v Sunningdale Joineries Ltd and ors* [2010] EWHC 2396 (TCC) at [5] to [13].

The "QS" is now a critical member of the project team and is likely to be involved with design and project management matters at all stages in the construction process.

2.35 At the outset of a project and during the period when the outline or concept design is being formulated the quantity surveyor will be often be engaged to provide costs information. One the key pieces of advice which the client will require from its professional team is their estimate of what the completed project will cost. That budget will encompass not just the construction costs but the professional fees, insurances, contingencies and incidental costs. Depending upon the design options which are open to the client, there may be a number of different costs models produced. If, as is usually the case, the project is budget-led (that is, designed to fit within a fixed cost), the design team and the quantity surveyor will work closely together to tailor the design to the constraints of the budget.

2.36 The quantity surveyor will generally be asked to prepare the detailed bill of quantities and schedules to enable contractors to tender for the work. When the tenders are submitted it may be called upon to advise as to whether certain parts of the tendered costings could be negotiated and will often be asked to take a lead role in any such negotiations. At this stage or even earlier the quantity surveyor may be asked to work with the design team or the design and build contractor in the process of value engineering (that is, attempting to produce a building or structure of the same or equivalent value for less cost). Depending upon the advisory role of an engineer or architect, the quantity surveyor may also be called upon to give more general advice, for example as to liquidated damages provisions or even the most suitable form of contract.[27] During the construction process, for works constructed under a traditional building contract, a quantity surveyor engaged by the employer will generally be required to assist the architect by advising the latter as to the value of work carried out for the purpose of providing interim and final payments and work to be carried out under proposed or actual variation instructions. Similarly, a quantity surveyor engaged by the contractor will be required to measure work carried out by the contractor and work carried out by subcontractors.

2.37 On projects constructed under design and build contracts the employer may not retain a quantity surveyor at all, whereas on projects undertaken by management contractors the employer will generally retain a quantity surveyor to shadow the functions of the management contractor's quantity surveyor.

2.38 Depending on the form of construction contract used, the role of the quantity surveyor may be prescribed in detail. At the completion of the works under a traditional construction contract a quantity surveyor engaged by the employer will generally be called upon to undertake a valuation of the contractor's final account. If the contractor has submitted a claim for loss and expense it will generally be required – in consultation with the employer's contract administrator – to advise the employer as to the value of that claim and to undertake such inquiries and negotiations as may be necessary.

2.39 Quantity surveyors are increasingly being called upon to perform "non-traditional" roles, for example as project managers or in relation to particular contracts, for example drafting the client's brief in relation to design and build contracts. The RICS Schedule

---

27 In a mirror image of these functions the quantity surveyor may be retained by a contractor in order to estimate the cost of works, price the bill of quantities, assist in the preparation of a tender and carry out negotiations. However most substantial contractors employ their own quantity surveyors and it would be relatively unusual for quantity surveyors to be independently engaged.

of Services categories 2 and 3 provide a useful indication of the range of services which quantity surveyors now offer.[28]

Professional regulation

2.40 Quantity surveyors are one branch of the wider family of surveyors which were, prior to 31 December 2000, divided into seven divisions by the Royal Institute of Chartered Surveyors ("RICS"), the others being commercial residential, building surveying, rural property, planning and development, land and hydrographic surveying and minerals surveying. After that date the seven divisions were replaced by 16 faculties (each representing a particular surveying skill, one of which is construction management) and in respect of each a member may subscribe up to four. There is no statutory regulation of qualification and practice, although in order to become a "chartered surveyor" the RICS requires certain qualifications and the passing of an "Assessment of Professional Competence". It operates a disciplinary code which makes professional indemnity insurance and continuing professional education compulsory and carries three levels of professional qualification, being Fellowship (FRICS), Membership (MRICS) and Associate (AssocRICS).[29] Fellows and Members are entitled to describe themselves as "Chartered" surveyors.

Quantity surveyors' contracts

2.41 As with architects and engineers, there is no legal requirement that quantity surveyors enter into any particular form of contract. The normal rules of the law of contract will apply to any agreement between a quantity surveyor and his client.

2.42 Thus, although it is prudent for such agreements to be in writing, or at least evidenced by an exchange of correspondence, agreements are often oral or partially in writing and partially oral. Similarly, although it is prudent for all terms of a contract to be agreed before the commencement of services it is common practice for the substantial terms to be agreed at the outset and for remaining matters to be implied according to custom and practice, or to be agreed subsequently.

2.43 The RICS Form of Agreement follows a familiar structure. The parties to the agreement are identified, as is the project. It is made clear that the Form of Agreement is to be read with the Form of Enquiry, Schedule of Services and Fee Offer which identify the work to be carried out and the payment to be made (quantity surveyors fees are no longer fixed to a particular scale and their remuneration is entirely a matter for commercial negotiation between the parties).[30] Condition 1.1 stipulates that the quantity surveyor will provide the services with reasonable care and skill. Condition 5 provides that he must carry

---

28 Of course, once quantity surveyors step outside their traditional roles and begin to advise on issues beyond matters of cost and value, the risks that a negligent error will lead to a substantial claim increase. Thus quantity surveyors who take on the role of advisers on design or who carry out the functions of a contract administrator may run the same risks as architects or engineers carrying out those functions. In the field of claims consultancy – which is dominated by quantity surveyors – advice which extends beyond the issues of cost and value into the sphere of legal advice may result in quantity surveyors running the same risks as lawyers.

29 Somewhat confusingly, "Members" were formerly known as "Associates" (ARICS), and "Associates" were formerly known as "Technical Members" (TechRICS).

30 In default of an agree sum or rate of remuneration, the Courts will imply into the contract a term that a reasonable sum be paid. However, it should be noted that this will not necessarily be the same sum as would have

professional indemnity insurance. Condition 11 provides for both use of a complaints procedure or adjudication in the event of a dispute whilst Condition 12 provides for voluntary arbitration.

2.44 Because of the narrow scope of his traditional functions it is usually less important that a quantity surveyor's contract of engagement spell out its precise role and responsibilities. These will generally be clear. That should not detract from the fact that where the quantity surveyor performs a designated function under a construction contract, the terms of that contract may determine the scope of its duties. In particular, quantity surveyors should be wary of provisions of the construction contract which may require the quantity surveyor to go beyond its usual tasks of valuation and the agreement of quantum and to empower him to make agreements as to liability.[31]

2.45 The RICS publishes a document entitled "Appointing a Quantity Surveyor" which consists of a guide for clients and surveyors, a standard form of enquiry setting out the details of the project together with the services which the quantity surveyor is to provide, followed by a fee offer which sets out the quantity surveyor's fees and lastly followed by the form of agreement which contains the terms of the appointment.

SECTION D: PROJECT MANAGERS

Origins

2.46 The increasing complexity and sophistication of construction projects has given birth to an entirely new type of construction professional. Often acting as the bridge between the employer and his design team and contractor commonly sits the "project manager".

2.47 The courts were initially reluctant to recognise project managers as a distinct professional group. In *Pride Valley Foods Ltd v Hall & Partners (Contract Management) Ltd (No. 1)*, HH Judge Toulmin CMG, QC viewed project managers through the prism of the services they provided, rather than as a distinct profession, stating that:

> There is an initial difficulty in accepting expert opinion evidence in relation to the duties of project managers. There is no chartered or professional institution of project managers nor a recognisable profession of project managers. In so far as it may be appropriate to accept expert evidence, the nature of the evidence that might be acceptable will depend on what the project manager has agreed to do. In some cases, the project manager will be the architect who will design the project and then, acting as project manager, supervise the contractor and the sub-contractors in carrying out the work. At the other end of the scale the project manager will supervise the work of the contractor and sub-contractors and ensure that the work is carried out in conformity with the design drawings. In these circumstances,

---

been customary, but will be one which takes into account all aspects of the engagement – see *Gwyther v Gaze* [1875] HBC (Fourth Edition) Vol. 2 p. 134.

31 See, for example, the facts of *John Laing Construction Ltd v County and District Properties Ltd* (1983) 23 BLR 1, where it was unsuccessfully argued that a provision enabling the employer's quantity surveyor to agree amounts payable to the contractor clothed him with authority to agree an absolute entitlement to those amounts, as opposed to merely agreeing the quantum of any entitlement. Had the argument succeeded the quantity surveyor may have found himself in breach of his duties to the employer.

the project manager will have no design function even to the extent of providing an outline specification.[32]

2.48 The following year in *Royal Brompton Hospital NHS Trust v Hammond and ors (No9)*,[33] HH Judge Lloyd QC noted that project management was still an emergent professional discipline, in which professional practices as such had not yet developed or become clearly discernible. In his view, the standard of care required of a project manager was likely to depend upon his particular terms of engagement and the demands of the particular project.

2.49 However, the increasing prevalence of project managers in large construction projects and the fact that the institutions representing surveyors and architects were happy to recognise project management as a distinct skill set (thus providing an ascertainable group of persons who could speak to standards in the profession) has meant that more recent judgments have treated project management as a distinct profession.[34]

Function

2.50 Project manager is a term of art appearing not just in industry usage but also within the framework of construction contracts themselves. Traditionally standard form contracts have accorded the administration role to "the architect", "the engineer", the "contract administrator" or the "employer's representative" (or "agent"), but it is noteworthy that the more recent (and successful) NEC3 and NEC4 forms accord the role to the "project manager". There is even a British Standard that "aims to help people and organizations achieve a desired outcome of a project efficiently and effectively, as well as to contribute to the learning within projects and so continually improve their organization's project management capability".[35]

2.51 The RICS defines project management as follows:

Project management is therefore concerned with defining what has to be accomplished, which is usually expressed in terms of technical performance (scope of works, which may include quality criteria, safety requirements, environmental considerations in addition to the technical performance criteria of the works), cost (budget) and time (programme or schedule). In simple terms project management does this by:

planning what needs to be done
implementing the plans
monitoring and controlling the project work; and
risk management.

The project manager leads and directs the project participants. He or she is accountable to the project sponsor for the project's successful completion (delivering the requirements on time and below budget)...

...a project manager is the person who is given responsibility for introducing change and is accountable to the project sponsor or project board for its successful accomplishment. The role

---

32 (2001) 76 Con LR 1 at 24.
33 [2002] EWHC 2037 (TCC).
34 See, for example, *Trustees of Ampleforth Abbey Trust v Turner & Townsend Project Management Ltd* [2012] EWHC 2137.
35 BS 6079-1:2010 Project management. Principles and guidelines for the management of projects.

of a project manager is therefore to lead and motivate the project participants to finish on time, within budget and to meet the requirements. This should result in satisfied clients.

2.52 This is a broad brief, as can be seen from the typical obligations commonly ascribed to project managers:

> The activities undertaken by the project manager typically include:
>
> - identifying needs and developing the client brief
> - leading and managing project teams
> - identifying and managing project risks
> - establishing communication and management protocols
> - managing the feasibility and strategy stages
> - establishing the project budget and project programme
> - coordinating legal and other regulatory consents
> - advising the selection/appointment of the project team
> - managing the integration and flow of design information
> - managing the preparation of design and construction programmes/schedules and critical path networks
> - advising on alternative procurement strategies
> - advising on risk management strategy
> - conducting tender evaluation and contractor selection
> - establishing time, cost, quality and function control benchmarks
> - controlling, monitoring and reporting on project progress; and
> - administering consultant appointments and construction contracts).[36]

Professional regulation

2.53 There is no distinct regulatory body representing project managers. In that sense the role of "project manager" may be thought to lack one of the essential attributes of a profession. That said, the position may be in transition. The Association of Project Management was founded in 1972 and was awarded a royal charter in 2017. It provides training and qualifications, as well as acting as a source of advice on matters of practice. It has a code of conduct and operates complaints procedure in respect of its members. It is one of a number of professional associations directed towards construction project management.[37] Moreover, given that almost all project managers are architects, engineers or quantity surveyors, who are members of the professional bodies regulating those professions, it would be wrong to think that there is no regulatory oversight of this group of professionals.

Project managers' contracts

2.54 Most of the larger professional associations representing architects and quantity surveyors produce standard form contracts for the engagement of project managers. Such contracts tend to list an extensive schedule of services to be provided by project managers covering all work stages in respect of a project.

---

36 RICS Guidance Note 107/2013: "Appointing a Project Manager".
37 For example, the International Project Management Association.

TYPES OF CONSTRUCTION PROFESSIONAL AND THEIR FUNCTIONS

SECTION E: BUILDING SURVEYORS

Origins

2.55 Building surveying has been recognised as a profession in the United Kingdom since the 1970s, and it is now one of the broadest areas of surveying practice, taking in projects from a domestic extension to a major retail development. Building surveyors provide professional advice on property and construction, completing detailed reports known as building surveys. They identify defects and advise on repair and maintenance options, but they can also be called upon to provide design services.

Function

2.56 Although building surveyors are commonly engaged for smaller projects where the client decides not to engage and architect building surveyors can be called upon to provide a broad range of construction and property related services. Typically, they may be appointed to:

- ensure projects are completed on budget and to schedule;
- advise clients on schemes and projects and determine requirements;
- prepare scheme designs with costings, programmes for completion of projects and specification of works;
- organise documents for tender and advise on appointing contractors, designers and procurement routes;
- determine the condition of existing buildings, identify and analyse defects, including proposals for repair;
- advise on energy efficiency, environmental impact and sustainable construction;
- instruct on the preservation/conservation of historic buildings;
- advise on the management and supervision of maintenance of buildings;
- deal with planning applications and advise on property legislation and building regulations;
- assess and design buildings to meet the needs of people with disabilities;
- instruct on construction design and management regulations;
- negotiate dilapidations (when there is a legal liability for a property's state of disrepair);
- carry out feasibility studies;
- advise on the health and safety aspects of buildings;
- advise on boundary and "right to light" disputes and party wall procedures;
- prepare insurance assessments and claims.

Professional regulation

2.57 Most building surveyors qualify as such and the most common route to qualification is by first taking a degree accredited by the RICS, the Chartered Institute of Building, the Association of Building Engineers and British Institute of Facilities Management, followed by two years' professional experience and on-the-job training.

Building surveyors' contracts

2.58 The very broad range of activities undertaken by building surveyors can be seen from the standard forms of appointment issued by the RICS.[38] There are eight different sets of services: "Construction", "Building and Measured Surveys", "Asset Management, Insurance", "Feasibility", "Property", "Landlord and Tenant" and "Miscellaneous". The "Construction" suite of services covers the entire gamut of the services generally provided by an architect acting as contract administrator.

SECTION F: OTHER CONSTRUCTION PROFESSIONALS

2.59 Modern construction projects can be distinguished from their ancestors by their complexity and by the diversification of the persons responsible for bringing the project to fruition. As is set out in the illustration below, the preparation of the initial scheme design may involve input from different consulting engineers as to matters such as geology, ground conditions, hydrology and drainage; planning and traffic flow consultants may be involved both in obtaining permissions and assisting with design; specialist consultants (not always engineers) may be called upon to advise as to specific elements of the design such as the foundations, the steelwork and the cladding; not only will the mechanical and electrical works require the input of a range of advisers, but information technology consultants, building management consultants, interior designers and landscape architects may all play a role.

2.60 Arguably a case can be made for the identification of IT professionals as a distinct class of construction professional. Very few modern buildings are not dependent, in some way, upon the successful design, installation and commissioning of IT systems (often referred to as "building management systems") the purpose of which is to make the building function efficiently.[39]

2.61 This complexity extends beyond the physical attributes of the works themselves. Whereas traditional buildings were the fruitful union of landowner and contractor, modern buildings have a plethora of parents: parties with interests in a modern construction project may include the landowner, the tenant, the developer, the funder and the sub-tenants. Traditionally an employer might have engaged a skilled person (comparable to the modern building surveyor) as a "clerk of works" to act as the employer's independent adviser as to the quality of what was being built, but in the modern construction project all of these persons might engage architects, engineers or quantity surveyors to advise them as to specific aspects of the works which affect them. Funders may engage someone with project management and quantity surveying skills to act as project monitors, advising the funder as to when and in what circumstances to release tranches of the loan. Disputes between the parties with economic interests in the building may bring about the engagement of other construction professionals. Architects, engineers and quantity surveyors may be engaged as adjudicators or arbitrators, or may be asked by the lawyers acting for the disputing parties to act as expert witnesses.

---

[38] "Building Surveyor Services" RICS 2008.
[39] In *Stephenson Blake (Holdings) Ltd v Streets Heaver Ltd* [2001] Lloyd's Rep PN 44 the court considered the usual duties which would be owed by an IT consultant to its client.

2.62 This process of specialisation holds out the prospect of the emergence of other distinct professions in the same way that project management has recently emerged. In particular, planning consultants play a very distinct and important role in the construction process and, notwithstanding the current judicial reluctance to view them as a profession,[40] the existence of the Royal Town Planners Institute (a chartered institute which seeks to maintain professional standards among planners) and the increasing volume of guidance as to standards and expectations from planning consultants suggest that recognition of the profession is not far off.

SECTION G: A TYPICAL CONSTRUCTION PROJECT

2.63 Construction projects vary from the smallest domestic extension to the construction of new railways or airports. Consequently, it is somewhat misleading to speak of a "typical construction project".

2.64 The following timeline is taken from a substantial commercial project carried out between 2004 and 2019 in the south of England. DV, a developer, has acquired a brownfield site near a major regional city with the express purpose of creating a distribution facility for AW, a major importer of foodstuffs. DV intends to sell the developed site to LD, an investment company, which will lease the property to AW.

Year 1: preliminary works

2.65 DV engages an architect and a planning consultant. The architect has primary responsibility for developing the outline scheme design and the planning consultant has responsibility for investigating planning and obtaining planning permission. A critical aspect of the architect's function is to work with consultants appointed by AW so that the design can be tailored to AW's needs. A second critical aspect is to advise DV on the appointment of a preliminary professional team. This will almost certainly include a civil and structural engineer and a mechanical and services engineer, and may well extend to specialist advisers (for example drainage consultants). The preliminary design team will in turn commission advice and reports from other consultants on aspects of the site (for example the load bearing properties of the ground) and the preliminary design (specialist elements of construction such as a design for the chilled areas). At some stage during this process, DV is likely to appoint a quantity surveyor (because obtaining an economic design will be critical to profitability) and a project manager (because the number of consultants and complexity of the work justifies that step).

Year 3: outline scheme design and application for planning permission

2.66 The consultants develop the outline scheme design in tandem with the application for planning permission. Once planning permission is obtained, the outline of the design (at least as it affects the general configuration of the buildings, roads and landscaping), will be effectively fixed. At this stage LD is likely to be taking a close interest in the

---

40 See *Elvanite Full Circle Ltd v AMEC Earth & Environmental (UK) Ltd* [2013] EWHC 1191 (TCC).

development of the design and the overall project scheme. LD and its funder will need to assess the viability of the project and the level of risk, both as to cost and time, presented by the anticipated construction programme. Depending on the nature of the building and the site it may retain its own construction professionals, for example a project management consultant and a civil and structural engineer. These persons will have a watching brief over the development of the design and may attend design team meetings along with representatives of AW.

Year 4: letting the contract

2.67 The design scheme is developed to the stage where there is a relatively prescriptive outline design. DV will then tender this design to prospective contractors, with the intention that the successful contractor will take entire responsibility for the design under a "design and build" contract. DV's consultants (in particular the architect, civil and structural engineer and mechanical and services engineer) must prepare this detailed preliminary design, which will take the form of drawings and specifications, so that the contractor will be able to design and build works which meet the client's needs (in this case, having had regard to AW's needs) and secondly so that the contractor is able to offer an economic price for the works (and the risk).

2.68 The quality of the design will thus directly impact on the client's long-term goal of obtaining a building which will match its contractual obligations to LD and AW and its short-term goal of transferring the risk of that design to a contractor for a price which will maximise its eventual profit. One of the key responsibilities for each of the consultants will be ensuring that his design is coordinated with the design of the others (and the architect and project manager will usually share overall responsibility for ensuring design coordination). A second responsibility is to ensure that the design is cost-efficient.

2.69 The project manager, or in his absence, the architect, will guide DV through the tender process. The evaluation of tenders involves both a consideration of the acceptability of the different contractors' proposals and a review of the costs which the different contractors may offer for different aspects of the works. It is relatively common at this stage for aspects of the design to change, either in response to individual suggestions by contractors or as part of a process of cost rationalisation. Each of the major designers appointed by DV will be involved in this process. DV's quantity surveyor and its project manager will play key roles. In particular, ultimate responsibility for ensuring that the selected contractor enters into a contract which meets DV's needs (insofar as practicable protecting DV from risk) will lie with the project manager.

Year 5: letters of intent, the building contract, novation

2.70 It is very common for works on construction projects to commence without a formal building contract. Thus, in this case construction works might be started by contractor C (through his specialist groundworks and piling sub-contractors) before the time when C's evaluators and lawyers have agreed formal terms with DV's project manager and DV's lawyers. In this situation, the legal relationship will generally be governed by a "letter of intent", a short-term contract which requires the contractor to work up to a certain level of expenditure. The responsibilities of DV's design team at this stage will often involve them working as if they were engaged by C, although no formal novation will usually take place until the Building Contract is agreed.

2.71 At the point at which C executes the Building Contract with DV, the design team (being at least the architect, civil and structural engineer and mechanical and services engineer) will generally have their contracts novated to C. Effectively they will enter into new contracts on the same terms as their engagements with DV, which replace their previous contracts. Under these novated contracts they will owe C the duties they owed to DV in respect of the work they carried out for DV. In this way C acquires rights against the consultants in respect of the design for which it is undertaking responsibility to DV. The design team will now work for C. They will transform the tender design into the construction design, ensuring at all times that it remains compliant with the tender design unless varied by DV.

2.72 This design team will undertake new responsibilities to DV and to LD and others. This will take the form of collateral warranties under which the design team agree to take responsibility for losses suffered by those persons as a result of any breach of duty under their contracts with C.

Year 5 to year 7: the construction period

2.73 During the construction period the design team transform the design as contained in the Building Contract (being the design prepared by them for the DV) into construction design. Persons unfamiliar with construction project might imagine that there is only one design and that the contractor proceeds on the basis of the plans and diagrams contained in the building contract, but reality is more complex. The construction design takes this tender design to new levels of detail and the process of producing the design must progress at a sufficient speed to ensure that the design is "frozen" (that is, final) in time for the contractor to construct the Works. Thus, the design team must programme and coordinate their design both with C and each other and they must sufficiently resource the Works so that they are able to respond quickly to queries raised by C and to the possible need for design changes.

2.74 The project manager, still engaged by DV, monitors the progress of the Works, attending progress meetings and looking out for DV's interests. Aside from its responsibility to monitor both progress and adherence to the terms of the Building Contract, the project manager is concerned with contract administration and particularly certification. C will be entitled to payment at various stages as the Works progress towards completion and it will fall to the project manager, assisted by the quantity surveyor, to assess progress and advise DV as the release of payments.

2.75 A further function falling to the project manager will be to operate the contractual machinery in respect of unplanned events. DV, perhaps on the prompting of AW, may wish to introduce a change into the design. This needs to be instructed as a variation and its consequences priced and negotiated. The Contractor may be entitled to an extension of time and the date for completion may have to be moved. Towards the end of the completion period works the Contractor may fail to meet the adjusted completion date and a certificate of non-completion may need to be served. When the Works are complete, it will be the responsibility of the project manager to issue a certificate of completion.

2.76 On a complex project such as this, construction professionals engaged by AW and LD may also be required to inspect progress, advise on critical aspects of the design and generally keep their clients informed. If there are aspects of the design or construction which are particularly important to AW it may instruct its architect and engineer to raise matters with the design team now working for the Contractor. As at completion of the works these construction professionals may be required to carry out their own inspections.

Year 7 to year 8: defects liability period

2.77 Almost all construction contracts require a period after completion, generally of a year, during which the contract is required to carry out small works which remained incomplete as at completion ("snagging") or to return to site to attend to defects which have become apparent. At this stage AW will be in possession of the Site and if problems emerge it will consult first with its own design team and then with DV, who may in turn require advice from the project manager and its design team. If problems are sufficiently substantial, construction professionals engaged by LD may be required to advise it.

Year 9 and subsequently: dispute resolution

2.78 If these problems can be resolved without difficulty the various construction professionals engaged on the Project will cease to be involved with it. However, if the problems cannot be resolved and develop into substantial disputes, not only will these construction professionals remain active in respect of the Works, but a new layer of activity by construction professionals may be added. Disputes may be referred to adjudication or to arbitration, and the adjudicators or arbitrators may themselves be construction professionals, or to litigation.[41] The parties to these disputes may be advised to obtain expert evidence, from other construction professionals, and there is even the possibility of experts being retained to advise as to the competence of the work carried out by the construction professionals originally engaged.

2.79 It is not unusual for construction disputes, often involving construction professionals, to be litigated or arbitrated many years after the Works themselves were complete.

---

41 Whether the mechanism for final resolution of disputes is arbitration or litigation will depend on the terms of the relevant contracts.

CHAPTER 3

The obligations of construction professionals

Section A: the contract	28
The function of the contract	28
When will a contract exist?	28
When will a contract be varied?	30
Section B: express terms	30
Ascertaining the meaning of express terms	30
Common types of problem	31
Express terms relating to services	31
Other express terms	33
Section C: implied terms	35
Section D: the tort of negligence	35
Section E: duty of care – personal injury	36
The general rule	36
Requirement of duty to the client	37
Section F: duty of care – physical damage	38
The general rule	38
No recovery for economic loss	38
No recovery where not just and equitable	39
Examples	40
Section G: duty of care to the client – economic loss	42
The general rule	42
The emergence of concurrent liability	42
Concurrent liability established	43
Section H: duty of care to third parties – economic loss	45
The general rule	45
Exceptions that failed	45
Assumption of responsibility	46
Section I: statutory duties	50
The Defective Premises Act 1972	50
The Construction (Design and Management) Regulations 2015	52
Section J: personal liability	54
Section K: remuneration and other matters	55
Remuneration	55
Disputes over remuneration	56
Plans, documents and copyright	57

SECTION A: THE CONTRACT

The function of the contract

3.1 The starting point (and often the ending point) for appreciating the obligations of a construction professional is its contract. The contract will define what it is that the construction professional is obliged to do, setting out the balance of its rights and obligations against the rights and obligations of the employer. As is discussed below, this is true even where the contract takes the form of the simplest written document or (as sometimes happens) no document at all.

3.2 The contracts under discussion here are contracts for the provision of services. The construction professional acts as an independent contractor, whose skills and labour are purchased on a project by project basis by the employer. We are not concerned with construction professionals who are engaged by developers or contractors as *employees*.

3.3 Construction professionals are generally engaged on the basis of detailed contracts, often in standard form or in standard form with bespoke amendments, but it is important to recognise that such agreements may not always be in place and, even where they are, they may not be clear as to responsibility for a particular issue. In those situations, any court or arbitrator trying to ascertain the extent of a professional's obligations will consider the meaning of the express terms and what can be taken to have been agreed by way of implied terms. Types of problem that occur frequently are discussed below.

3.4 A number of different terms are used to describe the construction professional's contract with its client. It can variously be referred to as "the contract", "the professional services agreement", "the appointment", "the engagement" and "the retainer". There is no practical distinction between this nomenclature. In law they are all contracts, and all these terms refer to the same set of mutual obligations.

3.5 On any particular construction project a construction professional may enter into a number of different contracts. On some projects, a construction professional may enter into more than one contract with the client (for example, for different work at different times) although this is unusual. More commonly the construction professional may enter into contracts with other persons. As a condition of the contract with the client, the construction professional may agree to enter into another contract with the client's funder or some person to whom the client intends to sell the completed project. In that agreement, the construction professional will undertake responsibility to that person for losses it might sustain as a result of a failure by the construction professional to carry out its work carefully. Such agreements are commonly referred to as collateral warranties or duty of care deeds.

3.6 Alternatively, the construction professional may agree that its contract with the client be novated to the design and build contractor. This means that a new contract comes into being between the construction professional and the design and build contractor with the latter stepping into the shoes of the original client.

When will a contract exist?

3.7 For a contact to exist, all that is required is evidence of an agreement that the construction professional will act (in some way) for the employer and will receive (in some way) benefit in return. The agreement is usually in writing, but it may be entirely oral: it is sufficient for the employer to ask the construction professional to carry out work and for the construction professional to agree to do so, or even, in some cases, simply to carry out the work. Under English law there is a predisposition to find that a contract exists in circumstances where one person carries out work for another, and the situation of construction professionals is no different.

3.8 However, the formation of a contract requires the presence of intent on the part of both parties to enter into a contract. This "intent" is objectively judged. What counts is what the parties do or say to each other rather than what they privately thought. The necessity for contractual intent means that the mere fact that work is carried out by a construction professional does not necessarily mean that a contract exists. Two situations must be distinguished.

3.9 Construction professionals frequently commence work for an employer prior to agreeing the terms of a formal contract. They may be agreed that their rights as against each other are to be subject to a final agreement. If no such final contract is agreed it is unlikely that the court or arbitrator will find that there was no contract at all. There will usually be a contract for some work to be provided in exchange for some payment (often a reasonable payment), but the other terms of that simple contract will be very limited. At its most extensive the contract will contain a term that the work carried out was carried out carefully, but no further obligations are likely to be imposed upon the construction professional.

3.10 This should be contrasted with the situation where the construction professional commences work without a written contract, but both parties are agreed on many if not most of the terms of that contract. If no final contract is entered into, it may be open to a court or an arbitrator to find that there was substantive agreement on a considerable number of terms, albeit not all of those contained in the document which was under negotiation. Much will depend upon the way in which the parties acted and what they said to each other.

3.11 There may be special circumstances where a construction professional carries out work for someone where neither the construction professional nor the person for whom the work is carried out intend (objectively) that a contract will come into being between them. These are almost always instances where the construction professional agrees to carry out work without payment or other benefit.[1] It will be a rare case in which some kind of contract does not come into being.[2]

3.12 Disputes of this nature are often about remuneration, rather than the quality of services or other obligations undertaken by the construction professional.[3]

3.13 A contract only gives rise to obligations between the persons who are the parties to it. A construction professional will almost always contract directly with the employer. It is possible for the employer to use an agent as the means of contracting with a construction professional, but this is very unusual. As a rule of thumb, there will always be a contract of some sort between a construction professional and its client.

3.14 There may be special situations where a construction professional carries out work under a contract with the client, which work is for the benefit of some third person. In those circumstances, it is possible that the true contract is with both, or that some other rule of

---

1 Consideration passing from the person receiving the benefit of the construction professional's work being an essential element for the existence of a contract between them.
2 See *Burgess and Anor v Lejonvarn* [2016] EWHC 40 (TCC); [2017] PNLR 25 (CA) for a case where the architect assisted a friend in the expectation that contract might be entered into but where there was no common intention to contract.
3 See Section K.

law allows the third person to take action against the construction professional,[4] but such circumstances are highly unusual.

When will a contract be varied?

3.15 Just as the making of a contract requires "intent" on the part of the employer and the construction professional, so there must be an objective intention shared by both parties if a contract is to be varied. The fact that a construction professional believes that it is no longer required to provide advice on a particular aspect of the works is unlikely to be sufficient to show that the contract has been varied to that effect. Something must have been said or done by the employer which clearly indicated that the employer wished to release the construction professional from that obligation. By the same token, the mere fact that the employer decides to add a new area of work to the construction project is not in itself a clear basis for asserting that the construction professional's contract has been varied so that it will now provide services in respect of that work. There has to be some evidence that both parties have agreed to change the contract in this way. Evidence of intention to vary may be obtained from the facts, but evidence of a unilateral intent will be insufficient.[5]

SECTION B: EXPRESS TERMS

Ascertaining the meaning of express terms

3.16 The express terms of a contract are the written (and sometimes oral) provisions of a contract setting out what the professional is required to do and how it is to do it. They are to be contrasted with what might be called the unwritten or unexpressed intentions of the parties as to those matters, which if they exist as contractual terms are said to be "implied".

3.17 A complex standard form contract such as the RIBA Standard Agreement contains a lengthy and detailed list of express terms. These will set out in detail what the architect is required to do (the services it is required to perform), when and in what circumstances that work is to be undertaken, the standard of performance, when and how much the architect is to be paid, what other obligations the architect might owe to the employer, how disagreements between the architect and the employer are to be resolved, and what limitations may exist on the rights of the employer to recover compensation from the architect.

3.18 Where the meaning of an express term is disputed the court or arbitrator will seek to ascertain that meaning by reference to the objective intention of the parties as demonstrated by the words they used and the context in which those words appear.[6]

---

4 For an example of a case where both contentions failed, see *Riva Properties Ltd and ors v Foster and Partners Ltd* [2017] EWHC 2574 (TCC).

5 It is not uncommon to find a clause in construction professionals' contracts which expressly prevent the construction professional from asserting that a variation has occurred unless the variation is agreed in writing. Such clauses will generally be effective but may not be a panacea as they may themselves be varied orally: see *MWB Business Exchange Centres Ltd v Rock Advertising Ltd* [2016] EWCA Civ 553 (at the time of writing an appeal on this point is pending in the Supreme Court.)

6 The question of how a court or an arbitrator goes about ascertaining the meaning of a contract (or more particularly, the meaning of word or phrase within a contract) is beyond the scope of this book. A useful shorthand guide can be found in *Arnold v Britton* [2015] UKSC 36.

3.19 Whilst the standard form contracts of most construction professionals have been drafted with the aim of avoiding obscurity (not least by employing terms which have a recognised meaning in the construction industry) it is not uncommon for professionals and employers to disagree as to the meaning of particular phrases and the scope of works to be undertaken.

Common types of problem

3.20 Although disputes over the meaning of a contract can arise in any number of ways, the following generic list covers the more common types of dispute:

3.21 *Disputes as to the true meaning of a word or phrase.* The lawyers responsible for the drafting of contracts between construction professionals and their clients strive to exclude any substantial possibility of disagreement over meaning by defining important words in clear terms. Most contracts will contain an extensive list of "definitions". However, that intent is not always successfully achieved and is of limited application if the parties choose to graft further documentation containing important words (which may not provide definitions of those words) onto standard terms. Thus, the engineer may have an obligation to review the design of "the groundworks" but the meaning of that term may not be fully and clearly set out, so that there is some doubt as to whether it includes the pavements of the car park.

3.22 *Conflicting provisions.* Carefully drafted contracts should be internally consistent, but particularly in complex contracts where the construction professional's services are contained in a number schedules, it may happen that there are apparently conflicting provisions. For example, under part of the contract dealing with the general description of the works the scope of the architect's responsibilities may appear to extend to designing all of the façade whereas, on one reading of the schedule of services attached, the architect may have no responsibility in respect of the design of the brise soleil.

3.23 *Omissions.* The contract may simply omit provisions which would explain to the parties what responsibility the construction professional has for a particular aspect of the works. Thus, for example, the engineer may be accorded responsibility for the design of the drainage in a development, but midway through the works the employer may discover that there are issues in relation to potential pollutants which need to be negotiated with the regulatory authorities. The engineer's contract may be entirely silent as to whether this is part of the engineer's scope of work.

Express terms relating to services

3.24 Most disagreements on the meaning of contracts arise in the context of the services which the professional is required to provide. Many construction professionals reach agreement on detailed contracts long after they have commenced work on the project. The terms which appear in the contracts (which may be terms advanced by the employer) may not correspond to the construction professional's understanding of the services which it has provided and has agreed to provide. This mismatch may be overlooked at the point at which the contract is executed. Alternatively, some issue may arise long after the services have been performed where the employer contends that something should have been done by the professional which was not done (possibly because no one gave any thought to it at the time) and the dispute concerns whether the wording of a particular part of the contract has the effect of imposing that responsibility upon the professional.

3.25 Areas of disagreement over services generally involve either responsibility for design which may have been carried out (or is thought to be the responsibility of others), or contracts which require the construction professional retrospectively to take responsibility for design or the coordination of design. Thus the civil and structural engineer may have overall responsibility for the design of a particular structure, but under the building contract a specialist sub-contractor may have design responsibility for a particular part of the building which it has agreed to provide. The employer and the professional may be in dispute as to whether, on a true construction of the express terms of the engineer's appointment, the engineer was entitled to leave certain aspects of the design to the specialist sub-contractor.[7]

3.26 In *Tesco Stores Ltd v The Norman Hitchcock Partnership Ltd*[8] the employer sued its architect for failing adequately to inspect the fire separation works which had been installed by contractors. The architect's appointment did not contain an express term which clearly extended to this responsibility and the employer's argument was rejected. Not only was there no express agreement that the architects would inspect this aspect of the works but the court found that the employer had not relied upon the architect to undertake this function.

3.27 Problems involving retrospectivity often arise in the context of novation. Here a construction professional will commence work for the employer but will subsequently have its contract novated to the design and build contractor (who is treated as having contracted with the construction professional from the start).[9] Because the employer and the design and build contractor may have had different interests (indeed sometimes conflicting interests) the design and build contractor may have a different appreciation of the extent to which a particular aspect of the services carried out by the construction professional on behalf of the employer was required to reduce risk in the design.[10]

3.28 Whilst it is important to distinguish disputes about the scope of services from disputes about the way in which services ought to be carried out, it is not always straightforward to identify the precise delineation between these two concepts. Many of the typical instances which are discussed in Chapter 4 concern the extent to which a particular service, expressly required by the contract, requires particular steps in order to perform it. Thus for example, the contract may require the architect to review the design of the specialist sub-contractor to ensure that it is generally in accordance with the contract documents. To what extent is the architect required to go in performing that obligation? Does it have to obtain advice from specialists in the field so that it can form an understanding as to whether or not the specialist sub-contractor's design will be generally in accordance with the contract documents?

---

7 See *Try Build Ltd v Invicta Leisure (Tennis) Ltd* (1997) 71 Con LR 141.
8 (1997) 56 Con LR 42.
9 Although technically this is a fresh contract. For a discussion of novation in this context see *J Jarvis & Sons Ltd v Castle Wharf Developments Ltd* [2001] EWCA Civ 19.
10 In *Blyth & Blyth Ltd v Carillion Construction Ltd* (2001) 79 Con LR 142 a question arose as to the point in time under the engineer's contract when the engineer would begin owing duties to the design and build contractor. The design and build contractor argued that, on its true meaning, the contract was fully retrospective so that the engineers were required to view the design and build contractor as their client from the start. The Scottish Court of Session disagreed. On the particular wording of the appointment the design and build contractor only became the client at the point of novation.

3.29 As will be seen, problems of this nature are often resolved not just by a consideration of the meaning of the particular words used in the appointment but also by their factual context and an ascertainment of the reasonable expectations of both the employer and the professional at the moment when the appointment was agreed.

Other express terms

3.30 Aside from the express terms which set out what it is that the construction professional has to do, the most important express term in an appointment is the term or terms setting out the standard to which that service is to be performed.

3.31 The almost universal standard is the standard of reasonable skill and care. The precise meaning of this phrase is discussed in Chapter 6, but it is important to note at this point that it is a dangerous over-simplification to say that a construction professional's obligation to his client is always to act carefully.[11]

3.32 In the first place, whilst the courts will frequently construe obligations in a construction professional's contract as obligations to act with reasonable skill and care, there is no necessity for the employer to limit the professional's obligation in this way. If the words of a contract are sufficiently clear, the employer may require the construction professional to bring about a particular result. This kind of absolute obligation may generally apply in terms of prohibitions (for example, preventing the construction professional from specifying a particular product or class of products) but it could conceivably apply to any aspect of the professional's services.

3.33 Secondly there are a class of ancillary obligations which may arise out of the construction professional's position as agent for the employer which impose absolute obligations (for example, not to enter into an agreement with a third party on behalf of the employer without the employer's authority).[12]

3.34 An important part of any contract between a professional and its employer is the allocation of risk. It is very common for the contracts of construction professionals to contain limitation of liability clauses. Such clauses are generally effective to "cap" the extent of the loss for which the construction professional can be liable in the event of a successful claim.

3.35 However, other broader protections are not uncommon. Many standard form appointments for architects and engineers contain clauses which require claims to be brought within a certain time (for example six years of the end of the defects liability period under the building contract), thus putting the risk of a late undiscovered claim against the professional on the shoulders of the employer.[13]

---

11 It is a common misapprehension that professional liability is to be equated with professional negligence and (more importantly) that the scope of a professional's obligation to his client to act carefully so that any task which a careful professional would undertake should be undertaken. This puts the cart before the horse. The scope of a professional's obligations is defined by the terms if its contract. The duty to act carefully attaches to the manner of performance of these responsibilities. See generally: *Midland Bank & Trust Co Ltd v Hett Stubbs & Kemp* [1979] Ch 384 at 435.

12 In contrast to other professionals, such as solicitors, the common law rules of agency and the equitable duties of fiduciaries rarely arise in respect of construction professionals: they rarely hold the client's property (especially money) and they rarely contract on behalf of the client. They are not excepted from the law in these fields: rather it will be a very rare case in which it is appropriate for the law to be applied.

13 See Chapter 6 where limitation is discussed.

3.36 Some contracts contain a "net contribution clause", a provision which limits the liability of the construction professional to that portion of the loss which it has caused which it is just and equitable for it to bear having regard to the responsibility of all the other persons who have caused the loss.[14] The default position in English law is that, where A and B, by their independent actions or omissions cause loss to C, C may generally recover *all* of the loss from whichever of A or B as it chooses to sue, regardless of their relative culpability for the loss. That party can then seek contribution from the other defendant. The fact that B has no resources to meet a claim provides no defence to A against C. That is A's risk. A net contribution clause alters this position. C can only recover from A that part of the loss which is A's share of overall responsibility. It thus shifts the risk to the employer that some other participant in the project who may be share responsibility in the loss will lack resources or will have a limitation defence by the time that the loss is discovered.

3.37 Further, as is discussed elsewhere,[15] the standard form appointments of construction professionals generally require that any dispute between the professional and the employer be resolved in arbitration rather than in litigation. This can present tactical advantages to the construction professionals.

3.38 It is partly for these reasons that most substantial employers will require construction professionals to agree to bespoke terms which strip out these and similar provisions and may often impose conditions that are favourable to the employer. Generally such contracts will be in the form of deeds, being contracts executed "under seal",[16] allowing contractual claims to be brought against the construction professional for a period of 12 years.[17] One provision which is particularly prevalent in such contracts is an obligation on the part of the construction professional to evidence that it has professional liability insurance at a particular level, to maintain that level of indemnity insurance for a number of years (often 12 years, corresponding with the applicable limitation period) and to provide evidence that the insurance is being maintained if called upon to do so at any time during that period.

3.39 The contract between a construction professional and its client is no different to any other contract when it comes to the general law regulating the operation of contracts. If the contract contains a particularly onerous term it is possible that its effect may excluded by operation of the Unfair Contacts Terms Act 1977 or (in the case of consumer contracts) the Consumer Rights Act 2015.[18] Similarly the contracts of construction professionals may be deemed to give rise to certain additional rights or obligations pursuant to the Contracts

---

14 See Chapter 6 where contribution is discussed.
15 See Chapter 9 where arbitration is discussed.
16 This terminology has survived the abolition of the requirement for deeds to be made using actual wax seals (and other special requirements) by s 1 of the Law of Property (Miscellaneous Provisions) Act 1989.
17 By virtue of s 8 of the Limitation Act 1980. Deeds are a "speciality", as opposed to a "simple contract".
18 In *Munckenbeck & Marshall v Michael Harold* [2005] EWHC 356 (TCC), the Court held that certain provisions of SFA/99 were onerous to the employer, who was a consumer, and because these had not been drawn to his attention when the appointment was entered into they were unfair and unenforceable under the Unfair Terms in Consumer Contracts Regulations 1999 (the predecessor Regulations to the Consumer Rights Act 2015). In *Oxford Architects Partnership v Cheltenham Ladies College* [2006] EWHC 3156 a provision of CE/95 which had the effect of shortening the limitation period was not struck down as unfair. Whilst both of these challenges concerned potential unfairness to the employer it is possible that, where an employer requires a construction professional to agree onerous terms favourable to the employer, the position would be reversed. Successful challenges to onerous terms in this area are less likely: such contracts will probably be individually negotiated and the courts may not view the negotiating positions of the professional and the employer as being in imbalance.

(Rights of Third Parties) Act 1999 and the Late Payment of Commercial Debts (Interest) Act 1998.

SECTION C: IMPLIED TERMS

3.40 At common law terms are implied into a contract when it necessary for some additional provision to regulate the rights and obligations of the parties because without that provision the contact either cannot function or because it is manifestly clear that the parties must have intended that their rights and obligations would be governed by this term and it was omitted from the express terms by oversight.[19]

3.41 At the risk of oversimplification, it can be said that the more detailed and exhaustive the express terms of a construction professional's contract, the less scope there is for the implication of terms. Because most construction professionals are engaged on standard terms and conditions or on bespoke terms which have been formulated by lawyers engaged by the employer, the implication of terms tends to play little part in disputes concerning the liability of construction professionals. However, arguments over implied terms do arise in situations where, for whatever reason, there is either an omission or a lack of clarity in the express terms.[20]

3.42 Moreover, whilst it is rare for a construction professional to be involved in a dispute over the disclosure of information confidential to its client, or a dispute with its client over a contract with a third party entered into by the construction professional on behalf of the client, it should be kept in mind that important obligations which arise as a consequence of the construction professional's status as the agent of its employer often arise as implied terms.

3.43 Implication for necessity or business efficacy is not the only route to implication.

3.44 Terms are also implied into the contracts of professionals by statute just as they are implied into other contracts. Thus section 13 of the Supply of Goods and Services Act 1982 implies an obligation into a contract to provide services that the services will be provided with reasonable skill and care.[21] Section 14 of the Act implies a term that the service will be provided within a reasonable time. Again, for the reasons which are set out above, these implied terms are unlikely to add anything of significance to the rights and responsibilities of the construction professional as against its employer.

SECTION D: THE TORT OF NEGLIGENCE

3.45 It is difficult to overstate the extent to which the construction professional's rights and obligations as against its client are governed by the construction professional's contract with the client. It is a rare case in which a court or arbitrator will need to look beyond that

---

19 This is shorthand for a much more nuanced test. In *Marks & Spencer Plc v BNP Paribas Securities Services Trust Co (Jersey) Ltd* [2015] UKSC 72 the Supreme Court discussed the multi-faceted test which is to be applied to implication in commercial contracts.

20 Thus, for example, in *Consarc Design Ltd v Hutch Investments Ltd* [2002] PNLR 31 there was a lack of clarity as to whether the architect's appointment had retrospective effect and if it did as to which obligations it provided. The court rejected the contention that there was an implied term that the appointment had the architect was obliged to provide warnings to the employer at a date prior to the appointment itself.

21 A different, but parallel regime exists for contracts with consumers under the Consumer Rights Act 2015.

contract. However, where it is necessary to consider legal obligations outside contract these are likely to be found in the tort of negligence.

3.46 In *Blyth v Birmingham Waterworks Co*[22], Alderson B described negligence as: "the omission to do something which a reasonable man, guided upon those considerations which ordinarily regulate the conduct of human affairs, would do, or doing something which a prudent and reasonable man would not do".

The mere fact of negligence is not sufficient to ground a cause of action, however: the claimant must show that the defendant owed it a *duty* not to be careless. In the tort of negligence A may owe a duty to B to take reasonable care in the way in which he goes about his business so as to avoid B from suffering harm.[23] This obligation can arise independently of any contractual relationship between A and B (although as described below, it arises in the context of some contractual relationships also).

3.47 The courts employ a number of different tools in order to decide whether in any particular case A is liable in tort for damage caused to B by its negligence. These tests primarily focus upon, first, the kind of harm suffered and, second, the kind of relationship between A and B.

3.48 As will be seen, in the context of construction professionals, the nature of the harm suffered by B has been decisive in determining whether a duty exists. Physical harm or damage to property will readily justify the existence of a duty of care. By contrast, financial loss which does not arise as the direct consequence of physical harm or damage to property, generally militates *against* the existence of such a duty. In those cases, something more will be required: something about the closeness of the relationship between A and B.

SECTION E: DUTY OF CARE – PERSONAL INJURY

The general rule

3.49 Construction professionals are in no different position to any other person when it comes to personal injury caused by their carelessness. Just as the driver of a car owes a tortious duty of care to passengers and other road users, or the surgeon owes a duty of care to his patient, the construction professional owes a duty to persons who might be injured by his carelessness to take reasonable care to safeguard them against such personal injury.

3.50 The key element in the existence of the duty is foreseeability: if A undertakes some activity which it is reasonably foreseeable might cause injury to B, A will generally owe B a tortious duty of care to safeguard him against injury caused by the careless carrying out of that activity.

3.51 Often the carelessness lies in an omission to act when a reasonably careful person would do something. In *Clay v AJ Crump & Sons Ltd*[24] demolition works had commenced on a site and the employer expressed concern to the architect as to the stability of a wall and

---

22 (1856) 11 Ex 781 at 784.
23 A claimant will be entitled to recover damages in the tort of negligence when three conditions are made out: (1) the defendant owes a duty of care to the claimant; (2) the defendant has acted in a way which is a breach of that duty of care and (3) the action of the claimant has caused the claimant to suffer *relevant* damage. Showing that a duty exists very much depends upon the closeness of the relationship between the claimant and the defendant and the kind of damage sustained. For a full analysis of the tort see *Charlesworth & Percy on Negligence*.
24 [1964] 1 QB 533.

the safety of persons working close to it. The architect promised to look into the matter. He consulted with the demolition contractor which assured the architect that the wall was safe. The architect did not carry out its own inspection. After completion of the demolition works in that area, the building contractor brought its workmen onto site. It carried out a cursory examination of the wall but did not notice its instability. The wall collapsed injuring one of the workmen. The Court of Appeal held that each of the architect, the demolition contractor and the building contractor owed the labourer a duty of care to take reasonable care to prevent him being injured.[25]

Requirement of duty to the client

3.52 It is important to note that the architect became subject to a duty of care because he had agreed to (or was obliged to) examine the stability of the wall. The mere fact that danger of personal injury exists in the environment in which a construction professional is working is not sufficient to give rise to a duty:[26] the injury must arise from the negligent carrying out of some task or responsibility of the construction professional.

3.53 In *Perrett v Collins*[27] the Court of Appeal explained the relationship between the task with which the defendant is charged and the existence of a duty of care:

> it has never been a requirement of the law of the tort of negligence that there be a particular antecedent relationship between the defendant and the plaintiff other than one that the plaintiff belongs to a class which the defendant contemplates or should contemplate would be affected by his conduct. Nor has it been a requirement that the defendant should inflict the injury upon the claimant ... In cases of personal injury, it suffices that the activity of the defendant has given rise to the situation which has caused the injury to the plaintiff. Where the defendant is involved in an activity which, if he is not careful, will create a foreseeable risk of personal injury to others, the defendant owes a duty of care to those others to act reasonably having regard to the existence of that risk.

and, later:

> Where the plaintiff belongs to a class which either is or ought to be within the contemplation of the defendant and the defendant by reason of his involvement in an activity which gives him a

---

25 Responsibility was apportioned 42% to the architect, 38% to the demolition contractor and 20% to the building contractor; although the architect was arguably less to blame than the demolition contractor, it was expected to exercise the greater degree of care in these circumstances.

26 Arguably the person injured must have been within the class of persons whom the construction professional's task or responsibility was reasonably intended to benefit. In *Harrison v Technical Sign Co Ltd* [2013] EWCA Civ 1569, a building surveyor was instructed by a landlord to inspect the awning of a shop which had been damaged by the landlord's workmen. The shop sign subsequently fell from the building and killed a passer-by. It was contended that the building surveyor had failed to notice that it was in a dangerous condition as in *Clay v Crump*, a duty of care was owed to anyone who might be injured by the failure to carry out the inspection with reasonable skill and care. The Court of Appeal found that there was no duty. The purpose of the inspection had been to advise the landlord in its dispute with the tenant. It was no part of the purpose to advise the landlord as to whether there was a danger to the public. The case was decided on its special facts, but it shows that where the construction professional does not cause the dangerous situation to arise (and its potential liability comes into being by its failure to prevent injury) it is not always enough to show a causal connection between the careless performance of a task and the ensuing injury: at least part of the purpose of the task or responsibility must have been to protect this class of person against harm arising in the way that it did. Older decisions along similar lines include *Clayton v Woodman & Son (Builders) Ltd* [1962] 2 QB 533 (CA) and *Oldschool and anor v Gleeson (Construction) Ltd and ors* (1977) 4 BLR 103 (QB).

27 [1998] 2 Lloyd's Rep 255 (CA).

measure of control over and responsibility for a situation which, if dangerous, will be liable to injure the plaintiff, the defendant is liable if as a result of his unreasonable lack of care he causes a situation to exist which does in fact cause the plaintiff injury.

3.54 As is discussed below, most construction professionals undertaking any kind of design role in respect of a building project will be subject to the provisions of the Construction (Design and Management) Regulations 2015. These regulations are specifically directed at promoting the preservation of the health and safety of any person affected by the project.[28] One consequence of the Regulations is to widen the class of persons within the contemplation of the construction professional to those who might suffer injury if the construction professional's work is not carried out carefully.

SECTION F: DUTY OF CARE – PHYSICAL DAMAGE

The general rule

3.55 Although the courts have been less eager to find that construction professionals owe a tortious duty of care to persons whose *property* suffers damage to property as a result of the carelessness of the construction professional than they have been to find a duty in respect of the claimant who suffers personal injury, the general rule is that provided it was foreseeable that damage would result from carelessness, a duty will exist.

3.56 The general rule is subject to two important constraints, both of which affect damage suffered by third parties.[29]

No recovery for economic loss

3.57 First, at least where the damage is sustained by a third party, damage must be physical damage: it cannot be economic damage. The distinction is intended to draw a line between physical harm and financial harm because whereas the former is readily ascertainable and identifiable as a known risk before the harm occurs, the latter is protean, potentially open-ended and often impossible to predict. In practice the distinction is not always straightforward. Thus, a wall that is cracked has suffered physical damage. A wall that is vulnerable to cracking has not suffered damage even though it may be about to crack. A wall which is cracked because it sits on the defective foundations designed by the engineer is physical damage. A wall which is cracked because the engineer carelessly designed that particular wall is economic damage: it is not damaged but is worth less than it should be.

3.58 The distinction arises in respect of construction professionals as a result of the development of the law of tort as to the duties of builders to third parties and, in particular, as a result of the development of the law concerning the rights of subsequent purchasers to sue local authorities and builders culminating in *Murphy v Brentwood District Council*.[30]

---

28 Regulation 8(1).
29 For the reasons which are set out in Section G, the duty of care owed to the construction professional's client will be no less extensive than its contractual duty.
30 [1991] 1 AC 398 (HL).

3.59 The courts have defined "pure economic loss" in terms of a distinction between the property which has been adversely affected by the carelessness of the construction professional. If an engineer working for Mr White carelessly designs the foundations of the house which Mr Blue subsequently purchases from Mr White, Mr Blue's loss is pure economic loss; there can be no claim against the engineer. This is so even though the engineer knew that Mr Blue was intending to purchase the house and would be harmed if the engineer carried out its work carelessly. Absent a contractual arrangement or some other relationship constituting an exception to the exclusionary rule, this is not the kind of loss which the engineer owes Mr Blue a duty to prevent. However, if those defective foundations cause cracking in an adjacent building owned by Mr Blue, that loss is physical damage; the engineer may be liable to Mr Blue in tort.

3.60 It should be noted that financial loss consequent upon physical damage is not classed as economic loss.[31] If the construction professional's breach of duty causes a fire which spreads to a neighbour's building the neighbour may recover the costs of having to repair the damage. The neighbour will also be entitled to recover damages for the loss of use of the building, provided these are not too remote.

No recovery where not just and equitable

3.61 The second constraint is that there may still be reasons why it would be unjust for a duty of care to be owed by a construction professional to a third party.

3.62 Thus in the example given above, the mere fact that the engineer's carelessness causes Mr Blue to suffer physical damage may not be sufficient; it must still be reasonable in all the circumstances for the engineer to owe Mr Blue a tortious duty of care. One of those circumstances may be the fact that Mr Blue could have discovered the defective nature of the design before he purchased from Mr White. Another may be that Mr Blue was a knowing participant in a contractual arrangement, involving Mr White, designed to prevent the engineer having a liability to Mr Blue.

3.63 Moreover not every third party who suffers damage to other property will be able to establish a duty of care.

3.64 The claimant must have a sufficient interest in the property to establish a duty.[32] In essence this means that the claimant must have some property right which the law recognises. The lawful occupier of a building who occupies under a lease or a licence will qualify. The potential purchaser of a building who has already expended monies in anticipation of purchase will not.

3.65 The act or omission of the professional which is said to have caused the damage must be one where the duty to the client and the duty to the third party coincide. The construction professional engaged to advise its client as to whether work is complete may not owe a duty to a third party to advise that the work is at risk of causing damage to some other structure. This is not least because the client may have engaged some other person to advise it against that kind of risk.

3.66 Other circumstances leading up to the breach may militate against the existence of a duty of care. If the third party has knowingly consented to a contractual regime where the

---

31 *Margarine Union GmbH v Cambay Prince Steamship Co Ltd* [1969] 1 QB 219.
32 *Leigh & Sillivan Ltd v Aliakmon Shipping Co Ltd* [1986] AC 785 (HL).

participants have purposefully limited the duties owed by the defendant (or have otherwise limited its responsibilities) it may be that it is not just and reasonable for a duty of care to exist.[33]

3.67 Finally the way in which the damage occurred may be so removed from the breach of duty that it is not just and equitable that a duty of care should exist. This is particularly relevant if the person who suffers the damage had the opportunity to discover the risk of damage posed by the breach of duty before that damage occurred.

Examples

3.68 The clearest examples of the operation of the law in the context of construction professionals are to be found in the Court of Appeal decisions in *Baxall Securities Ltd v Sheard Walshaw Partnership*[34] and *Bellefield Computer Services Ltd v E Turner & Sons Ltd*.[35]

3.69 In *Baxall* a developer appointed architects to design a warehouse. The siphonic roof drainage system was inadequately designed: it was unable to cope with extremely heavy rainfall. The warehouse flooded on two separate occasions, damaging goods owned by a company which had leased the warehouse. The company sued the architect for the damage. The architect denied that it owed the company a duty of care. It contended that the company had had the opportunity to discover the true state of the drainage before it took up the lease. Accordingly, the court should follow those parts of *Murphy* where it was held that it would be unfair to allow a duty of care in respect of other property when a subsequent purchaser had had the chance to discover the defects before purchase.[36]

3.70 The Judge at first instance accepted the argument in respect of one part of the design. The company could have discovered the absence of overflows (the fact that it did not was irrelevant). However, this was not enough to defeat the claim. The damage was also caused by the under-design of the system in that it was unable to cope with very heavy rainfall. This was not something which a careful purchaser of the lease, employing a building surveyor, would have been able to identify. Consequently, a duty of care was owed to the company and it was entitled to recover its losses. The Court of Appeal upheld this decision.

3.71 In *Bellefield* a builder negligently installed fire stopping in a dairy processing plant. Fire destroyed the building and its contents. The owners sued the builder who sought contribution from the architect who had provided the builder with some of the design. In order to succeed it was necessary for the builder to show that the architect owed the owners a duty of care.

3.72 At first instance the Judge held that (1) an architect may in appropriate circumstances, owe a duty of care in tort and be liable to a subsequent occupier of the building which the architect has designed and/or the construction of which he has supervised

---

33 See the discussion in *Biffa Waste Services Ltd v Machinenfabrik Ernst Hese GmbH* [2008] PNLR 17 (successfully appealed, but not on this point).
34 [2002] EWCA Civ 9.
35 [2002] EWCA Civ 1823.
36 Since *Donoghue v Stevenson* [1932] AC 562 it has been the law that a duty of care in respect of purchased or acquired property causing harm may be negated if the purchaser or acquirer has an adequate opportunity to identify the defect or danger prior to purchase or acquisition.

in respect of latent defects in the building of which there is no reasonable possibility of inspection and (2) whether a particular defect in a building comes within the scope of an architect's duty of care to a subsequent occupier will depend upon the original design and/or supervisory obligations of the architect in question. On the facts, the architects were not negligent because they had no duty to design the fire stopping and no duty to inspect it.

3.73 The Court of Appeal agreed. The Court identified guiding principles governing the tortious liability of construction professionals for physical damage to other property:

(i) An architect may, in appropriate circumstances, owe a duty of care in tort and be liable to a subsequent occupier of the building which the architect has designed and/or the construction of which he has supervised in respect of latent defects in the building of which there is no reasonable possibility of inspection.

(ii) The question whether a particular defect in a building comes within the scope of an architect's duty of care to a subsequent occupier will depend upon the original design and/or supervisory obligations of the architect in question. The architect will not owe a duty of care in respect of defects for which he never had any design or supervisory responsibility in the first place.

(iii) If a dangerous defect arises as the result of a negligent omission on the part of the architect, he cannot excuse himself from liability on the grounds that he delegated the duty of design of the relevant part of the building works, unless he obtains the permission of his employer to do so.

(iv) The detailed duties of an architect in relation to his design function depend upon the application of the general principles above stated to the particular facts of the case, including any special terms agreed. The precise ambit of such duties will usually depend upon expert evidence from members of the profession as to what a competent, experienced architect would do in the circumstances.

3.74 The condition that there should be no reasonable possibility of inspection may be justifiable on the basis of the general law, but its application may be limited on the facts. It is somewhat unsatisfactory the more gross the defect (the more it is detectable) the weaker the subsequent owner's claim against the construction professional. The condition has been justified on the basis that the failure to ascertain the defective nature of the property breaks the chain of causation, but on the basis of the current law of causation[37] that contention is probably misplaced. Certainly the condition has been criticised[38] and it may be that it is only in exceptional cases that it will be applied.

3.75 *Baxall* and *Bellefield* illustrate the fact-sensitive nature of the duty of care for physical damage in so far as that damage has been suffered by third parties. In those cases the responsibility of the construction professional to its client dovetailed readily with the responsibility of the construction professional to a subsequent purchaser. Just as it was plainly the professional's duty to his client to design a fully functioning building, so it was relatively straightforward to find that, subject to the issue of inspection, it was the professional's obligation to

---

37 For the act or omission of the defendant or a third party to break the chain of causation the intervening act must completely remove the causal potency of the original breach of duty. It is difficult to see this happening in all but exceptional cases.
38 See *Pearson Education Ltd v The Charter Partnership Ltd* [2007] BLR 324 (CA).

take reasonable care in designing so as not to cause damage to other property owned by the subsequent purchaser.

SECTION G: DUTY OF CARE TO THE CLIENT – ECONOMIC LOSS

The general rule

3.76 For the construction professional the most important non-contractual obligation to the client is the obligation to safeguard the client against "pure economic loss"[39] by not being negligent; an aspect of the so-called "concurrent liability" of the construction professional in contract and in negligence. If an employer contracts with a construction professional the latter will owe the employer a tortious duty of care which is co-extensive with its contractual duty. Just as the construction professional will be liable in contract for financial loss suffered as a result of its breach of duty, so it will be liable to the employer in negligence for economic loss.

The emergence of concurrent liability

3.77 The importance of this concurrent liability lies almost exclusively in the field of limitation.[40] Depending on the facts, the law of tort may provide a longer limitation period than under the law of contract. By the time that the employer discovers the loss or damage more than six years may have elapsed since the breach of contract. The cause of action in contract will be statute barred.[41] Yet it may still be open to the employer to sue in tort. The "damage" for which it sues may have commenced on a later date than the breach. The fact of the late discovery may enable the employer to benefit from the extended limitation period made available by Section 14A of the Limitation Act 1980. For this reason it is sometimes the case that claims against construction professionals by their employers succeed in tort where they would not have succeeded in contract and it is routine (even in cases where there is no risk of a limitation issue arising) for claimant employers to maintain a claim in tort in parallel to the claim in contract.

3.78 Because concurrent tortious liability adds so little to contractual liability, the development of the law in relation to the tortious liability of construction professionals has largely been in the context of their obligation to safeguard third parties against either physical damage or pure economic loss (see above). There was no watershed moment when the existence of the concurrent duty came to be recognised. Although there was support for the existence of such a duty as early as *Donoghue v Stevenson*,[42] the higher courts

---

39 The concept of pure economic loss is discussed below, but it generally refers to damage to the property which is the basis of the construction professional's contract.

40 See Chapter 6. Until relatively recently it was thought that claims in tort might be subject to more generous treatment under the law of remoteness. As at the date of this work the law appears to have moved away from that possibility, treating all concurrent tortious liability as being governed by the contractual rules. See *Wellesley Partners LLP v Withers LLP* [2015] EWCA Civ 1146.

41 The limitation period is 12 years if the contract was executed as a deed – see above.

42 [1932] AC 562 (HL).

were generally antipathetic until the 1970s. In *Bagot v Stevens Scanlan & Co Ltd*,[43] Lord Justice Diplock said that he:

> could see nothing in the relationship of architect and client which can be said to give rise to the kind of status obligation which arises from the origins of the common law in the case of master and servant, common carrier, innkeeper, bailor and bailee.

3.79 However, support for concurrent liability emerged in other contexts,[44] so that by the mid-1980s there were a number of cases in which the court assumed (or, at least, it was not in dispute) that a construction professional owed a tortious duty of care to his client which corresponded with its contractual duty.

3.80 Thus in *Pirelli General Cable Works v Oscar Faber & Partners*[45] the employer sued consulting engineers after a chimney was found to exhibit cracking. The claim was originally made in both contract and tort, but by the time it reached court the employer conceded that the contractual claim was statute barred. It was not contested that the engineers owed the employer a tortious duty to safeguard the employer from the economic loss of having to pay the costs of repairing the cracked chimney. The issue was whether the tortious claim was also statute barred.[46] The likelihood is that the court and the parties took the view that by entering into a contractual relationship the engineers were offering to provide the employers with advice in circumstances at least as favourable to the existence of a duty of care as the circumstances of the accountants in *Hedley Byrne v Heller & Partners*.[47]

3.81 Developments in the law of professional liability elsewhere swept construction professionals along in their wake. In *Henderson v Merrett Syndicates Ltd*[48] the House of Lords held that Names at Lloyds who contracted with managing agents were owed tortious duties of care in parallel with the agents' contractual obligations. The argument that the presence of contractual obligations acted to exclude a tortious duty of care was rejected. The reasoning in *Bagot v Stevens Scanlon* which distinguished between different types of engagement was also dismissed. This decision made it axiomatic that persons such as managing agents providing skilled services owed their employers tortious duties concurrent with their contractual obligations. There was (and is) no very plausible basis to distinguish construction professionals.[49]

Concurrent liability established

3.82 That said, it was only in 2011 that the Court of Appeal provided a definitive account of concurrent liability in relation to construction professionals, albeit very much by way

---

43 [1966] 1 QB 197 (QB).
44 For example *Esso Petroleum Co Ltd. v Mardon* [1976] QB 801 (CA) *and Midland Bank Trust Company Ltd v Hett, Stubbs & Kemp* [1979] Ch 384 (Ch).
45 [1983] 2 AC 1 (HL).
46 *Pirelli* is a difficult case which some commentators believe was wrongly decided on the issue of limitation. However, no-one has criticised the assumption that the engineers owed a concurrent duty of care.
47 This was the analysis of Lord Keith in *Murphy v Bentwood District Council* [1991] 1 AC 398 at 466f.
48 [1995] 2 AC 145 (HL).
49 Although it is right to note that there were a number of conflicting first instance decisions where there was disagreement as to whether the duty existed in the context of contractors who owed their clients design obligations – see *Storey v Charles Church Development Ltd* [1995] 73 Con LR 1 and *Payne v John Setchell Ltd* [2002] PNLR 7. As a result of *Robinson v P E Jones (Contractors) Ltd* [2011] EWCA Civ 9, it is likely that they do not (at least that was the view of Burnton LJ) – see below.

of distinguishing their position from the position of building contractors who were found to owe contractual duties only. *Robinson v PE Jones Ltd*[50] sets out a number of principles:

- In the relationship of contractor and employer the contract is the primary determinant of the obligations and remedies of the parties.
- The mere fact of contractual obligations is not generally sufficient to give rise to a duty of care on the part of the contractor to prevent economic loss.
- For such a duty to arise there must be an "assumption of responsibility" on the part of the contractor.
- One of the strong indicia of an assumption of responsibility is that the contractor is a professional person offering professional services; the employer is reliant upon his exercise of reasonable skill and care. The court's approach in these circumstances is shaped by authority stemming from *Hedley Byrne v Heller*.
- Outside the realm of professional retainers an assumption of responsibility must be proved according to the principles which govern such cases.

3.83 Although not cited by the Court of Appeal in *Robinson* this approach is consistent with the earlier Court of Appeal decision of *Barclays Bank plc v Fairclough Building Ltd*.[51] In that case the issue was whether and asbestos removal sub-contractor owed a tortious duty of care to the contractor. Beldam LJ said:

> A skilled contractor undertaking maintenance work to a building assumes a responsibility which invites reliance no less than the financial or other professional adviser does in undertaking his work. The nature of the responsibility is the same though it will differ in extent ... I would hold that [the sub-contractor] in performing the work ... owed a concurrent duty in tort to avoid causing economic loss by failing to exercise the care and skill of a competent contractor.

3.84 Admittedly there is no very satisfactory demarcation between design and workmanship. In the context of design and build contractors Ramsey J noted in *Harrison v Shepherd Homes Ltd*[52] that:

> as recognised in the judgment of May J in *Bellefield Construction v Turner*[53], there is a blurred borderline between design and the construction details needed to put the design into effect. As May LJ said "A carpenter's choice of a particular nail or screw is in a sense a design choice, yet very often the choice is left to the carpenter and the responsibility for making it merges with the carpenter's workmanship obligations". Where a contractor carries out both the design and construction of the works the intention is to avoid the blurred borderline, or sometimes gap, that may exist between the obligation to design and the obligation to construct. In such circumstances, a contractor responsible for both the design and construction who agrees that the works will be carried out in a proper and workmanlike manner, that is, with proper skill and care, cannot say that the works were constructed in a proper and workmanlike manner and avoid the same obligation applying to responsibility for the design. The works which include both the design and the construction must be carried out in a proper and workmanlike manner.

---

50 [2011] EWCA Civ 9.
51 (1996) 76 BLR 1 (CA).
52 [2011] EWHC 1811 (TCC) at [43].
53 [2002] EWCA Civ 1823 at [76]. The reference should be to *Bellefield Computer Services v Turner*, referred to above.

3.85 The fact of a blurred demarcation is more likely to move design and build contractors into the category of persons owing their clients a tortious duty of care, than to move other persons offering skilled services into the category of *Robinson* type contractors.

3.86 It follows that because the construction professional is in the business of offering skilled advice (something that helps define it as a construction professional) its relationship with its employer falls into that class of relationships governed by *Hedley Byrne* principles.[54] However, it is important to note that the assumption of responsibility critical for the existence of the *Hedley Byrne* duty arises from the rights and responsibilities allocated by the parties in their contract. This means the concurrent duty in tort is co-extensive with the contractual duty and certainly cannot be more onerous. The fact of the duty does not allow the parties to re-write the contract.[55]

SECTION H: DUTY OF CARE TO THIRD PARTIES – ECONOMIC LOSS

The general rule

3.87 As set out above, whereas a duty of care will readily be found to exist when the carelessness of A causes physical damage to B or B's property even where there is no contractual relationship between them, the courts are cautious in finding such a duty when the carelessness of A causes B to suffer loss of profit or a damage in the form of defective design of construction of property.[56] This is the operation of the general exclusionary rule against allowing claimants to recover against parties with whom they have no contract for "pure economic loss".

3.88 It is rule which is subject to exceptions and, as will be seen, the relationship between construction professionals and persons who might suffer loss as a result of their carelessness, but with whom they have no contract, often fits within one of these exceptions.

Exceptions that failed

3.89 Following *Murphy v Brentwood District Council*[57] and the related House of Lords decision in *D&F Estates v Church Commissioners for England*[58] valiant attempts were made by claimants to describe the property which had suffered harm as "other property". This "complex structure theory" posited that, in a complicated building, negligence in respect of one aspect of the design or construction causing damage in another distinct part of the structure was not negligence causing economic loss, but negligence causing physical damage to a different part of a complex structure. Its origins lie in speculation in both cases that such circumstances may allow for a duty of care permitting recovery for the economic consequences of one part of a building causing harm to another part. These attempts have

---

54 As will be seen, where there is no contractual relationship, but the circumstances are very close to those of a contract, the courts have been willing to find that such a tortious duty arises. See for a recent example, *Burgess v Lejonvarn* [2016] EWHC 40.

55 See *Greater Nottingham Co-operative Society Ltd v Cementation Piping & Foundations Ltd* [1989] QB 71 (CA).

56 The rule against the recovery of pure economic loss goes back to the nineteenth century and in particular *Cattle v Stockton Waterworks Company* [1875] LR 10 (DC).

57 [1991] 1 AC 398 (HL).

58 [1989] AC 177 (HL).

been consistently and robustly rejected[59] and it is suggested that notwithstanding the pedigree of its origins, the theory is difficult to justify and if applied likely to lead to inconsistent decisions.

3.90 In *Murphy* the House of Lords considered a further possible exception to the exclusionary rule. Lord Bridge considered that if the claimant's building contained a defect which rendered it dangerous to neighbouring property, but that it had not yet caused damage to that property, the claimant might sue the person responsible for carelessly causing or permitting the defect for "the cost of obviating the danger, whether by repair or demolition, so far as that cost is necessarily incurred in order to protect himself from potential liability to third parties".[60] The potential exception is not without its difficulties and its narrow factual scope means that it is seldom likely to be a relevant route of recovery for subsequent property owners. However, it demonstrates some of the conceptual difficulties involved in defining the recovery of pure economic loss by reference to the distinction between the property in respect of which the construction professional was engaged and "other property".

Assumption of responsibility

3.91 The only substantial and effective exception to the exclusionary rule is that a construction professional may be liable to a third party for economic loss resulting from the carelessness of the construction professional where it has assumed a responsibility to the third party within the meaning of that phrase as applied in *Hedley Byrne & Co v Heller & Partners*.[61]

3.92 In *Hedley Byrne* the House of Lords decided that bankers owed a duty of care to a third party which relied upon references they had prepared. The references constituted a negligent misstatement made in circumstances in which the bankers knew that the third party would rely upon them.[62] Lord Reid stated that in order for a defendant to be liable to a claimant which had relied upon the defendant's statements the defendant must have undertaken some responsibility to the claimant. In *Henderson v Merrett Syndicates Ltd (No. 1)*[63] the principle enunciated in *Hedley Byrne* in relation to negligent misstatements was extended to the negligent performance of services. Lord Goff stated[64] that the principle underlying *Hedley Byrne* was an assumption of responsibility by the defendant which "rests upon a relationship between the parties which may be general or specific to the particular transaction, and which may or may not be contractual in nature". Accordingly,

---

59 *Warner v Basildon Development Corporation* [1990] 7 Const LJ 146; *Payne v John Setchell Ltd* [2002] PNLR 7; *Broster v Galliard Docklands Ltd* [2011] EWHC 1722 (TCC) (where Akenhead J questioned whether the theory "still has a material part to play in the law of negligence relating to buildings and structures"). There is one decision which may be consistent with the theory, *A Jacobs v Moreton and Partners* (1995) 72 BLR 92, but there were very unusual facts in this case: the consultant engineers had not designed the house which was damaged but were retained to design repairs to the house which consisted of a new and independent foundation system.
60 [1991] 1 AC 398 at 475.
61 [1964] AC 465.
62 The claim failed because the bankers had given the references "without responsibility".
63 [1995] 2 AC 145.
64 At 180.

a person who assumes responsibility to another person to perform a service and fails to do so with reasonable care may be liable for the loss suffered by that other person in relying on the assumption of responsibility.

3.93 In determining whether an assumption of responsibility has occurred an objective test must be employed. As Lord Steyn stated in *Williams v Natural Life Foods Ltd*:[65]

> An objective test means that the primary focus must be on things said or done by the defendant or on his behalf in dealings with the plaintiff. Obviously, the impact of what a defendant says or does must be judged in the light of the relevant contextual scene. Subject to this qualification the primary focus must be on exchanges (in which term I include statements and conduct) which cross the line between the defendant and the plaintiff.

The claimant must rely upon the misstatement and that reliance must be reasonable in the sense that it must have been the kind of statement upon which that kind of defendant would reasonably rely (opinions offered voluntarily in situations where, objectively, they were not intended to be relied upon will not give rise to a duty).

3.94 The facts may negate the existence of a duty. In a situation involving more than two parties, although in principle a party may assume responsibility to more than one person in respect of the same activity, the way the parties have structured their relationships may be inconsistent with an assumption of responsibility and that will make the existence of a duty less likely.

3.95 However, it is important to note that the concept of "assumption of responsibility" is not without its critics and in determining whether a duty of care in respect of economic loss exits the courts will view assumption of responsibility as just one way of assessing a particular set of facts. A more inclusive test is generally applied, testing the outcome by reference to three different approaches. The other two tests are:

3.96 *The threefold test:* The threefold test in *Caparo Industries plc v Dickman*[66] asks whether: (a) loss to the claimant was a reasonably foreseeable consequence of what the defendant did or failed to do; (b) the relationship between the parties was one of sufficient proximity; and (c) in all the circumstances it would be fair, just and reasonable to impose a duty of care on the defendant towards the claimant. The test is well established and has been applied in many cases; and

3.97 The incremental test:

> When confronted by a novel situation the court does not ... consider [the stages of the threefold test] in isolation. It does so by comparison with established categories of negligence to see whether the facts amount to no more than a small extension of a situation already covered by authority, or whether a finding of the existence of a duty of care would effect a significant extension to the law of negligence.[67]

3.98 As can be seen, there is a tension between the test of assumption of responsibility and the threefold test. The latter is more general in its application. Cases concerned with the liability of construction professionals to third parties for pure economic loss demonstrate different balances between these two tests.

---

65 [1998] 1 WLR 830 at 835.
66 [1990] 2 AC 605 at 617–618.
67 Per Phillips LJ in *Reeman v Department of Transport* [1997] PNLR 618 at 625.

3.99 At one end of the spectrum are cases such as *IBA v EMI and BICC*[68] and *Hunt v Optima (Cambridge) Ltd.*[69]

3.100 In *IBA* the employers engaged main contractors to design and erect a television mast. The main contractors engaged sub-contractors to carry out the works. The sub-contractors were asked to and did give the employers an assurance that the mast would not oscillate dangerously. In fact the design was negligent and the mast collapsed. The House of Lords held that the sub-contractors were in breach of a duty of care to the employers and were liable for the consequences of the collapse.

3.101 In *Hunt* an architect was engaged by a developer to administer the development of a block of flats. Part of the architect's duties was to provide completion certificates to purchasers certifying that their flats had been constructed in accordance with the building contract. There were defects in the construction and the architect was found to be liable to the purchasers for the costs of putting these right. The liability arose not by reason of the architect's negligent supervision of the works, but by reason of the misstatement in the certificates.[70]

3.102 Similarly the civil and structural engineers in *Payne v John Setchell*[71] were liable to the purchasers by reason of their negligent certification of the raft foundations.

3.103 The further away the facts from situations of this kind, and the closer one gets to the other end of the spectrum, the less likely it is that a duty of care will exist.

3.104 In the absence of express statements an assumption of responsibility can exist by the mere performance of duties, but it is often hard for a claimant to establish that the performance of the duty was anything more than the construction professional comply with its contractual obligations to its client. In these circumstances the threefold test is likely to yield a more accurate analysis of the legal relationship between the construction professional and the third party.

3.105 In contrast to the facts of *Payne v John Setchell*, in *Preston v Torfaen Borough Council*[72] the consultant engineers who had carried out the inspections upon which the foundation design was based gave no certificates to the purchasers. It was accepted for the purposes of the preliminary issue that the engineers held themselves out as experts in respect of that kind of investigation, knew that the local authority (which was building the houses) would rely upon its advice, knew that the local authority would sell the houses to persons like the claimants and knew that if their work had been careless the purchasers were likely to suffer economic loss. The Court of Appeal held that the engineers owed no duty of care. There was no assumption of responsibility and on the application of the threefold test no duty arose.

3.106 Not only may the facts fail to disclose the ingredients of an assumption of responsibility, but they may militate against one.

3.107 In *Sainsbury's Supermarket Ltd v Condek Holdings Ltd and ors*,[73] the owners of a supermarket sued consulting engineers alleging that they had acted in breach of their

---

[68] (1980) 14 BLR 1.
[69] [2014] EWCA Civ 714.
[70] On appeal the architect succeeded against many of the claimants on the basis that they had purchased their flats before they had received the certificates and consequently cannot have relied upon them.
[71] [2002] PNLR 7.
[72] (1994) 65 BLR 1 (CA).
[73] [2014] BLR 574.

tortious duty of care in the negligent design of the car park. The engineers applied to strike out the action as disclosing no reasonable cause of action. The Court granted the application. There was no arguable duty of care. The parties had structured their relationships so as to exclude the possibility of the engineers owing a tortious duty of care to the owners. Stuart-Smith J cited a passage from Lord Goff's speech in *Henderson* which indicated that a sub-contractor would not generally owe a duty of care to an employer because "there is generally no assumption of responsibility by the sub-contractor or supplier direct to the building owner, the parties having so structured their relationship that it is inconsistent with any such assumption of responsibility".[74] The Judge added:

> Although it may be said that this passage is strictly obiter, it accurately reflects the present state of the law which, in this respect, has been settled for a considerable time. Nor is it necessary that the contractual framework be structured in any particular way or with any particular degree of complexity. That is because it is for the person asserting the existence of a duty of care to prove the existence of the special relationship that is a necessary prerequisite to it; and a contractual structure which does not show any unusual or particular features indicating a special relationship of proximity between the employer and the independent subcontractor will be taken as conforming to the norm, as explained by Lord Goff.
>
> Lord Goff was referring expressly to the sub-contractor who provides work or materials in the construction of a building, and not to professionals such as NRM. However, once it is recognised that a claim in tort against a building professional who is not in a contractual relationship with the Claimant requires the existence of a special relationship of proximity as the foundation for the existence of a relevant duty of care, Lord Goff's statement of principle can be seen to be equally applicable to sub-contracted design professionals as to sub-contracted suppliers of work and materials. The scope for an employer to identify facts which justify the conclusion that there has been an assumption of responsibility towards it by a design professional differs from that in respect of a building sub-contractor supplying work and materials: but the underlying principle remains the same.[75]

3.108 In the usual situation, the construction professional either contracts with the employer or provides the employer with a collateral warranty. Other persons who are sued by the employer in respect of defective construction works can generally sue the construction professional for contribution on the basis that both are liable to the employer for the same damage. However, where a contractor suffers loss which has been caused or contributed to by an error on the part of a construction professional, the contractor may have no contractual claim and is unlikely to be able to claim contribution. In these circumstances contractors have attempted to claim against construction professionals in tort but those attempts have almost entirely failed.[76]

3.109 In *Pacific Associates v Baxter*[77] the contractor contended that the certifying engineers owed it a duty of care in the process of certification (alleging that as a result of negligent under-certification of interim certificates it had suffered loss). The Court of Appeal rejected that argument. Such a duty was inconsistent with the contractual structure and, in particular on the facts of this case, the contractual remedy made available to the contractor to challenge the correctness of certificates by arbitration.

---

74 [1995] 2 AC 145 at 195.
75 At paragraphs 45 and 46. For an earlier decision along similar lines see *Ove Arup & Partners International Ltd v Mirant Asia-Pacific Construction (Hong Kong) Ltd (No. 2)* [2004] EWHC 1750.
76 In *Michael Salliss & Co Ltd v Calil and William F Newman & Associates* (1987) 13 Con LR 68, the Court expressed a measure of support for the existence of a duty of care owed by an architect to a contractor in respect of unfair certification. However, the correctness of this decision was doubted in *Pacific Associates v Baxter*.
77 [1990] 1 QB 993 (CA).

3.110 For a construction professional engaged by an employer to owe a tortious duty of care to a contractor something more is required than the mere fact that the contractor may suffer loss if the construction professional acts carelessly: there must be some act or omission on the part of the construction professional which takes the case out of the norm and creates a responsibility on the part of the construction professional.

3.111 This is illustrated by *J Jarvis & Sons Ltd v Castle Wharf Developments Ltd*.[78] A contractor commenced work at a project intending to enter into a design and build contract. The work was halted after the planning authority asserted that it was in breach of planning permission. The contractor sued the project manager which, it alleged, had made negligent misstatements to the effect that the contractor's design, based on modified employer's requirements, would be in conformity with planning requirements. The project manager asserted that it owed no duty of care. The claim failed on causation (the contractor could not show it had relied on any representation by the project managers) but the Court of Appeal distinguished between loss caused by reliance upon representations made by the project manager prior to any contractual nexus between the contractor and the employer, when a *Hedley Byrne* duty might exist, and representations made by the project manager acting as the employers agent in the context of a design and build contract, when there would be no such duty.

3.112 As to the former, Gibson LJ said:

> There is no reason in principle why the professional agent of the employer cannot become liable to a contractor for negligent misstatements made by the agent to a contractor to induce the contractor to tender, if the contractor relies on those misstatements. But whether a duty of care in fact arises in any given situation must depend on all the circumstances, including in particular the terms of what was said to the contractor.[79]

SECTION I: STATUTORY DUTIES

The Defective Premises Act 1972

3.113 The Defective Premises Act 1972 provides a remedy to any person who either engages someone else to design and/or construct a dwelling or who acquires an interest in that dwelling. Section 1(1) of the Act states:

(1) A person taking on work for or in connection with the provision of a dwelling (whether the dwelling is provided by the erection or by the conversion or enlargement of a building) owes a duty –

    (a) if the dwelling is provided to the order of any person, to that person; and
    (b) without prejudice to paragraph (a) above, to every person who acquires an interest (whether legal or equitable) in the dwelling; to see that the work which he takes on is done in a workmanlike or, as the case may be, professional manner, with proper materials and so that as regards that work the dwelling will be fit for habitation when completed.

---

[78] [2001] EWCA Civ 19.
[79] At paragraph 53.

3.114 It will be noted that the cause of action has a number of essential ingredients: (1) the property must be a "dwelling"; (2) the dwelling must have been unfit for habitation when completed and (3) the fact that it is unfit for habitation must be because the defendant has failed to carry out its work in workmanlike or professional manner with proper materials. The cause of action is open to anyone who causes the dwelling to be constructed or who acquires an equitable or legal interest in it.

3.115 "Dwelling" is a building intended to be used as a residence. In a block of flats each separate apartment is a "dwelling" and although the common parts (stairwell, halls, car parks etc.) are not part of the dwelling they are none the less brought within the scope of the duty because they are "in connection with the provision of a dwelling".[80] Plainly defects to parts of a development within which a dwelling is situated will only be material insofar as they render the dwelling unfit for habitation.

3.116 "Unfit for habitation" plainly imports a high threshold of defect. Defects which are merely irritating or which substantially devalue a property may not be sufficient: they have to be defects which either present a risk to the health of the inhabitants or which so detract from the amenity of the property that no one could be expected to live there. That said, provided that at completion the defect crosses the threshold of presenting sufficient risk of becoming unfit, it does not matter that the property has not actually become unfit.

3.117 In *Andrews v Schooling & Ors*,[81] Balcombe LJ observed:

> [i]f, when the work is completed, the dwelling is without some essential attribute – e.g. a roof or damp course – it may well be unfit for habitation even if the problems resulting from the lack of that attribute have not then become patent. A house without a roof is unfit for habitation even though it does not rain until some months after the house has been completed.

3.118 There is some doubt about the proper meaning of the last ingredient of the cause of action. In *Harrison v Shepherd Homes Ltd*,[82] Ramsey J expressed the view that the Act gave rise to a threefold duty: to see that the work is done, first, in a workmanlike or professional manner; secondly, with proper materials and, thirdly, so that as regards that work the dwelling will be fit for habitation when completed. Breach of any one of these obligations would give rise to a cause of action. However, he felt bound by the Court of Appeal's decision in *Alexander v Mercouris*[83] that there was a single duty, to see that work was done in a workmanlike or professional manner with proper materials with the particular result that the dwelling will be fit for habitation when completed. On this basis, the lack of skill or use of inadequate materials must render the property unfit. The first interpretation has much to recommend it. In particular the Law Commission plainly intended a much broader remedy. However, pending the matter returning the Court of Appeal, the law is as stated in *Alexander*.

3.119 The words "a person taking on work for or in connection with a dwelling" are very broad. They will catch any construction professional undertaking professional services in relation to a property intended for residential use. The fact that the construction

---

80 See *Rendlesham Estates plc and ors v Barr Ltd* [2015] EWHC 3968 at [33]–[54].
81 [1991] 1 WLR 783 (CA).
82 [2011] EWHC 1811 (TCC).
83 [1979] 1 WLR 1270 (CA), followed by *Thompson v Clive Alexander & Partners* (1992) 59 BLR 77.

professional's error may not be the primary cause of the unfitness is probably immaterial. Provided its error was a material cause, the action will be made out.[84] Thus the architect who negligently fails to notice that the contractor has omitted to install the damp proof course correctly can be sued for the entire cost of remedying the defect. The engineer who negligently fails to warn that the foundation design might be inadequate can be liable for the entire cost of remedial piling works even though the superstructure (designed and built by others) was also substantially defective.

The Construction (Design and Management) Regulations 2015

3.120 The Construction (Design and Management) Regulations (the CDM Regulations) first came into force in 1994 and were amended in 2007. The 2015 Regulations update and improve their effect. In essence they seek to inculcate a safety culture into the business of construction projects. The Health and Safety Executive summarises their intent as follows:[85]

CDM aims to improve health and safety in the industry by helping you to:

- sensibly plan the work so the risks involved are managed from start to finish
- have the right people for the right job at the right time
- cooperate and coordinate your work with others
- have the right information about the risks and how they are being managed
- communicate this information effectively to those who need to know
- consult and engage with workers about the risks and how they are being managed

3.121 Whereas the 2007 CDM Regulations focused upon the role of the CDM Coordinator, the 2015 Regulations focus on the client, the "principal designer" and the "principal contractor".

3.122 Regulations 4 and 5 set out the client's duty to make suitable arrangements for managing a project and maintaining and reviewing these arrangements throughout, so the project is carried out in a way that manages the health and safety risks. For projects involving more than one contractor, these regulations require the client to appoint a principal designer and a principal contractor and make sure they carry out their duties. Most of the client's duties can be delegated to the principal designer and principal contractor. Some projects must be notified to the HSE.

3.123 Regulation 8 sets out a number of requirements on anyone working on a project with certain responsibilities.[86] They relate to appointing designers and contractors, the need for cooperation between dutyholders, reporting anything likely to endanger health and safety, and ensuring information and instruction is understandable.

3.124 Regulations 9 and 10 set out the duties placed on designers. It is therefore particularly relevant to construction professionals. These duties include the duty to eliminate, reduce or control foreseeable health and safety risks through the design process, such as

---

84 There is a limitation period of six years from the date of completion: s 9 of the Limitation Act 1980.
85 HSE website.
86 The Regulations distinguish between clients, principal designers, principal contractors, designers, contractors and workers.

those that may arise during construction work or in maintaining and using the building once it is built:

(i) A designer must not commence work in relation to a project unless satisfied that the client is aware of the duties owed by the client under these Regulations.

(ii) When preparing or modifying a design the designer must take into account the general principles of prevention and any pre-construction information to eliminate, so far as is reasonably practicable, foreseeable risks to the health or safety of any person – (a) carrying out or liable to be affected by construction work; (b) maintaining or cleaning a structure; or (c) using a structure designed as a workplace.

(iii) If it is not possible to eliminate these risks, the designer must, so far as is reasonably practicable – (a) take steps to reduce or, if that is not possible, control the risks through the subsequent design process; (b) provide information about those risks to the principal designer; and (c) ensure appropriate information is included in the health and safety file. (4) A designer must take all reasonable steps to provide, with the design, sufficient information about the design, construction or maintenance of the structure, to adequately assist the client, other designers and contractors to comply with their duties under these Regulations.

3.125 Regulation 11 sets out the duties a principal designer has during the pre-construction phase. The principal designer may be anyone with control over the pre-construction phase of the project, typically the architect or project manager. The duties include requirements to plan, manage, monitor and coordinate health and safety during this phase and to liaise with the principal contractor in providing information relevant for the planning, management and monitoring of the construction phase.

3.126 Regulation 12 sets out the duties on either the principal designer or principal contractor for the preparation, review, revision and updating of construction phase plans and health and safety files. This role generally falls to the principal contractor, working with the assistance of the principal designer but it is possible for the principal designer to take on the role. However, even where this happens, it remains the principal contractor's responsibility to plan and manage the construction works. Regulation 13 sets out the principal contractor's duties during the construction phase. The main duty is to plan, manage, monitor and coordinate health and safety during this phase. Other duties include making sure suitable site inductions and welfare facilities are provided.

3.127 Breach of the CDM Regulations can result in criminal prosecution. For the purpose of civil actions, whilst the regulations create no cause of action in themselves[87] they undoubtedly colour actions brought against construction professionals in contract and in tort. Failure to follow the Regulations or aspects of the HSE guidance which accompanies them is likely to be treated as evidence of carelessness, requiring rebuttal by the construction professional. An act or omission by a construction professional causing personal injury which is also a breach of the Regulations is likely to make it difficult (if not impossible) to sustain an argument that no duty of care was owed.

---

87 Recognising this, many bespoke contracts expressly provide that the construction professional must comply with its obligations under the Regulations. This gives the employer a potential cause of action for damages should the professional breach its duties under the Regulations.

SECTION J: PERSONAL LIABILITY

3.128 The foregoing discussion of the obligations of construction professionals makes no distinction between the construction professional and the company or other entity for whom that person works. This is distinction in determining whether the construction professional, or the entity for which that person works, will be liable in the event that a breach of contract or tortious duty is made out. As is explained in Chapter 8, construction professionals who work for companies or partnerships are usually insured in their own right just as their company or partnership is insured. Moreover in unusual circumstances persons who work as construction professionals for companies or firms offering the services of construction professionals can be sued personally. As a general guide, the position is as follows.

3.129 Where a construction professional contracts with a client on his or her own behalf and/or carries on business on his or her own account as a sole trader, any liability to the client will be the liability of that person.

3.130 If Ms Mary Black runs her own architectural practice trading as "Mary Black Associates" she will have a personal liability in the event that a client is able to prove breach of duty causing loss. Even if the negligence was committed by someone working for Mary Black (an assistant), Ms Black will be liable as the contracting party. The same situation pertains in the law of tort. If negligent design by Mary Black Associates causes injury to a third party, Mary Black herself will be liable. If the negligence was committed by an employee, Mary Black will be vicariously liable for that employee's torts.

3.131 Where a construction professional contracts with a client on behalf of a partnership and/or carries on business a partner, any liability to the client will be the liability of that person and all the other partners in partnership at the date when the breach of duty occurred.

3.132 If Ms White is a partner in White & Brown, a firm of consulting engineers, she and Ms Brown will have a personal liability in the event that a client is able to prove breach of duty causing loss. Even if the error was made by Ms Brown, in relation to a project in which Ms White was not involved, Ms White will be jointly and severally liable as a partner, both in contract and in tort. Again, the partnership will be liable for the negligent acts of employees.

3.133 Where a construction professional contracts with a client on behalf of a limited liability partnership and/or carries on business a partner in a limited liability partnership, any liability to the client will be the liability of the limited liability partnership. Indeed this is the principal reason that limited liability partnerships exist.

3.134 If Ms White is a member in White & Brown LLP, a firm of consulting engineers, the LLP will be liable in the event that a client is able to prove breach of duty causing loss, but (as a general rule) no liability will attach to Ms White or Ms Brown personally. The LLP will be responsible for the negligent acts or omissions of its employees.

3.135 Where a construction professional contracts with a client on behalf of a limited company the position is the same as for a limited liability partnership. The company, like an LLP, is a separate legal person. It is the company which takes (legal) responsibility for the acts of its directors or employees. This is so even if the company is in reality merely a vehicle for a single construction professional.

3.136 If Ms Mary Black runs her own architectural practice through "Mary Black Ltd" (and contracts with clients in that way) the company will have liability for breaches of contract and she will generally have no personal liability in the event that a client is able to prove breach of duty causing loss.

3.137 This statement of the general position is applicable in most claims against construction professionals. However, possible exceptions exist.

3.138 In the first place (and obviously) it matters little that a construction professional trades through a limited company or an LLP if he or she *in fact* contracted with the client in a personal capacity. The courts will judge the identity of contracting parties objectively, looking at the words used and their meaning against the background of facts known to both parties at the moment when the contract is entered into. If Mary Black sent the employer a copy of an RIBA standard form agreement which merely identified Mary Black as the architect, the fact that invoices are subsequently sent and paid in the name of Mary Black Ltd may not alter the fact that the employer's contract was with Mary Black.

3.139 Second, in cases of tortious liability the employer will generally only be liable for the negligent acts or omissions of the employee where the negligent act or omission was undertaken in the course of the employer's business. Thus if an unqualified engineer employed by White & Brown LLP gives some structural advice to a neighbour, it may be difficult for the neighbour to sue White & Brown LLP if the advice was negligent and loss was caused.

3.140 More unusually, there may be circumstances where although the client contracts with the limited company or LLP, the client places such particular trust and confidence in the construction professional working for the limited company or LLP that that person owes a tortious duty of care which may exist in parallel to the client's contractual rights. Such cases are naturally rare and would involve conduct on the part of the construction professional which made it just and equitable that such an unusual duty should exist,[88] but they could exist in cases where the construction professional's advice is central to the project and both the client and the construction professional objectively regard their relationship as being with each other, in reality disregarding the role of the limited company or LLP.

3.141 That situation must be distinguished from the much more common position of a limited company or LLP offering a named person to carry out services as a construction professional or indeed situations where a named person is identified in a contract between other parties as the individual who will carry out a certain role (for example, certifier). This kind of specific identification may limit the ability of the company or LLP in how it performs the tasks which the named individual is required to perform (substantial performance by someone else is very likely to be a breach of contract)[89] but it does not generally give rise to any kind of assumption of responsibility on the part of the individual so named.

SECTION K: REMUNERATION AND OTHER MATTERS

Remuneration

3.142 The Construction Professional's contract with its client will set out the terms upon which the construction professional is entitled to be paid. In the absence of such terms, a term will be implied that a construction professional is entitled to reasonable payment for the work done.[90] Disputes about remuneration are common, but they rarely concern the

---

[88] See *Williams v Natural Life Health Foods Ltd* [1998] 1 WLR 830 (HL). Some professionals (not only in construction) seek to include clauses in their contracts that prevent clients from suing employees directly.
[89] See, for example, *St Modwen Developments (Edmonton) Ltd v Tesco Stores Ltd* [2006] EWHC 3177.
[90] Section 15 of the Supply of Goods and Services Act 1982.

contractual mechanisms for payment. In the usual case, these are clear and the dispute concerns whether the construction professional has done something (or failed to do something) so that the client is entitled to refuse or reduce payment.

3.143 Before considering those situations, it is helpful to make a number of points about entitlement to remuneration.

3.144 The contract determines entitlement. This means that the construction professional which agrees to carry out work for a specific sum (for example, a proportion of the building costs) is not entitled to seek additional payment merely because the work turns out to be more onerous than anyone expected.[91]

3.145 If the client wants the contract administrator to carry out more work or different work to that which is the subject of the contract, the construction professional may agree to carry out that work on such further terms as to payment as it may negotiate. If it carries out the work in the expectation of being paid, but without an agreement, the court will usually regard the parties as having agreed that a reasonable sum would be paid.

3.146 The contract not only governs what the construction professional is entitled to be paid, it governs when the construction professional is entitled to be paid. If, as is now usually the case, the construction professional's contract falls under the jurisdiction of the Housing Grants, Construction and Regeneration Act 1996,[92] there will be an implied entitlement to stage payments and to the protection afforded by the "pay less" provisions, which require an employer to give particular notice of any decision to pay less than the sum due on the particular interim payment.

3.147 The contract will also govern the construction professional's entitlement to expenses. No statutory term will be implied to deal with expenses in the event that the contract is silent on this matter, but in the event that the construction professional has incurred costs for the benefit of the client in circumstances where the client knew and encouraged the construction professional to incur those costs, it is likely that the costs would be recoverable under a form of claim known as "quantum meruit".[93]

Disputes over remuneration

3.148 Most disputes are concerned with alleged failures on the part of the construction professional adequately to carry out its work.

3.149 In such cases the fact that the construction professional may have carried out its services poorly (and in breach of its obligation to employ reasonable skill and care) is not, in itself, a reason why the construction professional is not entitled to payment.

3.150 The employer may be entitled to refuse or reduce payment if and only if the construction professional's breach of duty has caused the employer financial loss. Moreover, the extent of the reduction cannot be greater than the financial loss actually sustained.

3.151 Subject to compliance with the provisions of the Housing Grants, Construction and Regeneration Act 1996 the employer may refuse to make payment without having to prove its loss (and the extent of the loss may be unclear), but it runs the risk of adjudication

---

91 See *Gilbert & Partners v Knight* [1968] 2 All ER 248.
92 See Chapter 9.
93 This is a restitutionary remedy whereby a party which has enabled another to become unjustly enriched, in the absence of a contract, can recover its loss to the extent of the enrichment.

or court proceedings by the construction professional to recover its fees. In that adjudication or in those proceedings the employer will need to establish, on the balance of probabilities, that there was both a breach of duty and that that breach of duty caused it loss which was equal to or more than the fees withheld.

Plans, documents and copyright

3.152 A well-drafted contract will determine whether ownership of plans, drawings and other intellectual property created by the construction professional remains the property of the construction professional or becomes the property of the client. Most agreements provide that the intellectual property in plans or drawings remains the property of the construction professional, but may be used by the client.

3.153 In the absence of such contractual provisions the general rule is that such documents become the property of the client once the construction professional has been paid.[94] Each case will depend upon the court's view of the underlying intentions of the parties

3.154 The position in respect of other intellectual property which is not constituted by plans or designs, such as computer programmes produced for the purpose of carrying out work for the client, is more difficult. The likelihood is that it remains the property of the construction professional.

3.155 Documents provided to the construction professional when the construction professional is working for the client which are provided in the course of that project generally belong to the client. A similar rule applies to documents generated by the construction professional (such as letters or emails).[95]

---

94 *Cala Homes (South) Ltd v Alfred McAlpine Homes East Ltd* [1995] FSR 818.
95 *Leicestershire CC v Michael Faraday and Partners Ltd* [1941] 2 KB 205 (CA).

CHAPTER 4

The standard of reasonable skill and care

Section A: the *Bolam* test	59
Section B: exceptions to the *Bolam* test	62
Section C: establishing the standard	65
Section D: codes of practice	67
Section E: the band of reasonable actions	68
Section F: outcome, not methodology	69
Section G: seniority, resources and price	71
Section H: delegation	72
Section I: duty to review	74
Section J: skill and care not enough	76

SECTION A: THE *BOLAM* TEST

4.1 As set out above in Chapter 2, the contracts of construction professionals almost always require the professional to carry out services with reasonable skill and care. The precise formulation of this contractual standard may vary (and that can introduce subtle variations in the standard) but the practical consequences of such variation are nugatory. In each case, the way in which the professional acted will be measured against the way in which a court or arbitrator thinks that a hypothetical competent construction professional of the relevant discipline would have acted placed in exactly the same situation. This is known as "the *Bolam* test".

4.2 Because it is a purely legal construct, the test has a number of technical facets and it is impossible adequately to discuss these without citing passages from the leading authorities.

4.3 However, before discussing precisely how the court or arbitrator applies the *Bolam* test, it is worth noting the reasons for the application of this test and its origins. Both in contract and in tort, the servant or agent might be liable to the master or employer for the failure to exercise the care and skill reasonably to be expected of that servant or agent.[1] The principle that a person who provides services calling for particular skill undertakes to exercise that degree of skill and care as is to be expected of a reasonably competent

---

1 That there is a large class of cases in which the foundation of the action springs out of privity of contract between the parties, but in which, nevertheless, the remedy for the breach, or non-performance, is indifferently either assumpsit or case upon tort, is not disputed. Such are actions against attorneys, surgeons, and other professional men, for want of competent skill or proper care in the service they undertake to render: actions against common carriers, against ship owners on bills of lading, against bailees of different descriptions: and numerous other instances occur in which the action is brought in tort or contract at the election of the plaintiff.
(Tindal CJ in *Boorman v Brown* (1842) 3 QB 511 at 525 (KB))

professional can be traced back to at least the mid-nineteenth century. In *Harmer v Cornelius*,[2] Willes J said:

> When a skilled labourer, artizan, or artist is employed, there is on his part an implied warranty that he is of skill reasonably competent to the task he undertakes – Spondes peritiam artis. Thus, if an apothecary, a watch-maker or an attorney be employed for reward, they each impliedly undertake to possess and exercise reasonable skill in their several arts. The public profession of an art is a representation and undertaking to all the world that the professor possesses the requisite ability and skill.

4.4 With the growing recognition of the skilled services provided by the professions, and particularly the medical profession, the courts established a benchmark for professional competence. In his direction to the jury in *Bolam v Friern Hospital Management Committee*,[3] McNair J stated:

> where you get a situation which involves the use of some special skill or competence, then the test whether there has been negligence or not is not the test of the man on top of the Clapham omnibus, because he has not got this special skill. The test is the standard of the ordinary skilled man exercising and professing to have that special skill. A man need not possess the highest expert skill at the risk of being found negligent. It is well established law that it is sufficient if he exercises the ordinary skill of an ordinary competent man exercising that particular art ... a mere personal belief that a particular technique is best is no defence unless that belief is based on reasonable grounds ... a doctor is not negligent if he is acting in accordance with ... a practice (accepted as proper by a responsible body of medical men skilled in that particular art), merely because there is a body of opinion that takes a contrary view.

4.5 The *Bolam* test posits a comparison between the actions of the defendant and the hypothetical actions of the ordinarily competent member of the profession. It is the foundation for assessing the competence of all defendants in "professional negligence" actions and it applies to construction professionals just as much as it does to doctors, solicitors or accountants. Lord Diplock said in the lawyers' case of *Saif Ali v Sydney Mitchell & Co.*:[4]

> No matter what profession it may be, the common law does not impose on those who practice it any liability for damage resulting from what in the result turn out to have been errors of judgment, unless the error was such as no reasonably well-informed and competent member of that profession could have made.

4.6 In the context of an architect, this duty to use reasonable care and skill was explained in *Nye Saunders & Partners v Alan E Bristow*[5] by Stephen Browne LJ in the following terms:

> Where there is a conflict as to whether he has discharged that duty, the courts approach the matter on the basis of considering whether there was evidence that at the time a responsible body of architects would have taken the view that the way in which the subject of enquiry had carried out his duties was an appropriate way of carrying out the duty, and would not hold him guilty of negligence merely because there was a body of competent professional opinion which held that he was at fault.

---

2 (1858) 5 CB NS 236 at 246 (QB).
3 [1957] 1 WLR 582 at 567 (QB).
4 [1980] AC 198 at 220 (HL).
5 (1987) 37 BLR 92 at 103 (CA).

4.7 Many Judges dealing with cases involving construction professionals have cited the standard as it was described by Windeyer J in respect of architects in the Australian case of *Voli v Inglewood Shire Council*:[6]

> An architect undertaking any work in the way of his profession accepts the ordinary liabilities of any man who follows a skilled calling. He is bound to exercise due care, skill and diligence. He is not required to have an extraordinary degree of skill or the highest professional attainments. But he must bring to the task he undertakes the competence and skill that is usual among architects practising their profession. And he must use due care. If he fails in these matters and the person who employed him thereby suffers damage, he is liable to that person. This liability can be said to arise either from a breach of his contract or in tort.[7]

4.8 It is important to note that, subject to the narrow class of exception set out below, the mere fact that one body of professional opinion is superior to another (better reasoned, better evidenced and so forth) does not mean that practitioners following the inferior practice are negligent. In the medical case of *Maynard v West Midlands Regional Health Authority*,[8] Lord Scarman said:[9]

> I have to say that the judge's preference for one body of distinguished professional opinion to another also professionally distinguished is not sufficient to establish negligence in a practitioner whose action have received the seal of approval of those whose opinions, truthfully expressed, honestly held, were not preferred. If this was the real reason for the judge's finding, he erred in law even though elsewhere in his judgment he stated the law correctly. For in the realm of diagnosis and treatment negligence is not established by preferring one respectable body of opinion to another. The failure to exercise the ordinary skill of a doctor (in the appropriate speciality, if he be a specialist) is necessary.

4.9 The mere fact that a substantial number of practitioners in a profession hold to a particular view is not sufficient to make it "respectable". In *Bolitho v City and Hackney Health Authority*[10] Lord Browne-Wilkinson observed:

> The Court is not bound to hold that a defendant doctor escapes liability for negligent treatment or diagnosis just because he leads evidence from a number of medical experts who are genuinely of opinion that the defendant's treatment or diagnosis accorded with sound medical practice ... the court has to be satisfied that the exponents of the body of opinion relied upon can demonstrate that such opinion has a logical basis. In particular in cases involving, as they so often do, the weighing of risks against benefits, the judge before accepting a body of opinion as being responsible, reasonable or respectable, will need to be satisfied that, in forming their

---

6 [1963] ALR 657.
7 [1963] ALR 657 at 661. For an early statement in similar vein relating to engineers see *Greaves & Co (Contractors) Ltd v Baynham Meikle & Partners* [1975] 1 WLR 1095 (CA) at 1101, per Lord Denning MR. In the more recent case of *Cooperative Group Ltd v John Allen Ltd* [2010] EWHC 2300 (TCC), Ramsey J. noted at paragraph 150:

> The applicable standard against which to judge the performance of [the defendant engineer] is that of an ordinarily competent structural and civil engineer, exercising ordinary care and skill. If a structural and civil engineer acts in accordance with the practice and views of a reasonable body of other structural and civil engineers, they do not act negligently: *Bolam v Friern* [1957] 1 WLR 582 at 587 to 588. The standard is that of "the reasonable average as Bingham LJ said in his dissenting judgment in *Eckersley v Binnie & Partners* (1988) 18 ConLR 1 at 79 to 80. The court has to be careful to judge the conduct by what was known at the time and not with the wisdom of hindsight.

8 [1984] 1 WLR 634.
9 At 639.
10 [1998] AC 232 at 241 (HL).

views, the experts have directed their minds to the question of comparative risks and benefits and have reached a defensible conclusion on the matter.

4.10 What constitutes a logical basis will depend upon the facts. It is suggested that it may be helpful to apply a two-stage test. First, that the persons constituting "the body of opinion" must have turn their mind to the comparative risks and benefits relating to the matter. Second as a result of that process, a defensible conclusion must have been reached: the opinion must be internally consistent on its face; it must make cogent sense as a whole, such that no part of the opinion contradicts with another and does not fly in the face of proven extrinsic facts relevant to the matter; it should not ignore or controvert known facts or advances in knowledge.[11]

4.11 Part of the business of exercising the care and skill of the ordinarily competent member of the profession is keeping abreast of the learning and standards of the profession, but in so doing the professional is required to keep up, not to be a pioneer. In *Eckersley and ors v Binnie and ors*,[12] Bingham LJ said:

> a professional man should command the corpus of knowledge which forms part of the professional equipment of the ordinary member of his profession. He should not lag behind other ordinarily assiduous and intelligent members of his profession in knowledge of new advances, discoveries and developments in his field. He should be alert to the hazards and risks inherent in any professional task he undertakes to the extent that other ordinarily competent members of the profession would be alert. He must bring to any professional task he undertakes no less expertise, skill and care than other ordinarily competent members would bring but need bring no more. The standard is that of the reasonable average. The law does not require of a professional man that he be a paragon combining the qualities of polymath and prophet.[13]

4.12 Where the construction professional is engaged with an unusual problem or employs new "cutting edge" techniques the standard of reasonable skill and care still applies (the case is not taken out of the *Bolam* arena merely because no, or no substantial, body of construction professionals in that discipline have faced that problem or used that technique before). Rather the court does its best to establish the standard which would be adopted by the ordinarily competent member of the profession faced with that problem or using that technique. Very often this will require enhanced caution and/or warnings to the client.[14]

SECTION B: EXCEPTIONS TO THE *BOLAM* TEST

4.13 Whilst the *Bolam* test is almost always applicable whenever there is a question as to whether a construction professional acted with reasonable skill and care, it should be noted that it is not *universally* applicable. There are limited exceptions.

---

11 Taken from the decision of the Singapore Court of Appeal in *Khoo James v Gunapathy* [2000] SGCA 25.
12 (1988) 18 Con LR 1 (CA).
13 Ibid at 80. This passage was quoted with approval by the Court of Appeal in *Michael Hyde & Associates Ltd v JD Williams & Co Ltd* [2001] PNLR 233 at [24], per Ward LJ (with whom Sedley and Nourse LJJ agreed).

14 For architects to use untried, or relatively untried, materials or techniques cannot in itself be wrong, as otherwise the construction industry can never make any progress. I think, however, that architects venturing into the untried or little tested would be wise to warn their clients specifically of what they are doing and to obtain their express approval.
(His Honour Judge Newey QC in *Victoria University of Manchester v Hugh Wilson & Lewis Womersley and Pochin (Contractors) Ltd* (1984) 2 Con LR 43 at 73 (QB))

4.14 The only true exception is where the careless act or omission may be so plain and/or so unrelated to any issue of practice that the court or arbitrator has no need to inquire into the standards of the profession and whether the construction professional acted "in accordance with ... a practice (accepted as proper by a responsible body of [construction professionals] skilled in that particular art)". Thus an architect who accidentally reverses the setting out drawings so that north is south and vice versa commits an act of carelessness which *can* be judged by the man on the Clapham omnibus. The engineer who fails to make any inquiry into the load bearing properties of the ground at a new and untested site will find it difficult to persuade a court or arbitrator that it needs to consider the general practice of engineers. It must be kept in mind that the *Bolam* test originates not because some special rule of law or evidence applies to actions against professional persons, but because it is usually the case that the act or omission which is the subject of the claim against them is concerned with the exercise of an esoteric or special skill. There is nothing esoteric or special about elementary blunders.

4.15 Secondly, and indicated above, it is not unknown for members of a profession either to adopt or (in substantial numbers) to follow a practice which, on analysis, lacks rigour or is objectively unsatisfactory. The *Bolam* test refers to the standard of conduct "accepted as proper by a *responsible* body of medical men skilled in that particular art". The fact that a substantial body of reputable and skilled construction professionals follow a particular practice is generally sufficient to show that following the practice is not negligent, but this is not automatically so. As set out above, a body of opinion must be objectively reasonable. The inference of competence may be negated if it can be shown that the practice was objectively unsound for reasons which should have been apparent to anyone practising in the field had they thought about it sufficiently carefully.[15] On analysis this is not a true exception at all. For the reasons given above, an unreasonable body of opinion simply falls outside the scope of "a practice (accepted as proper by a responsible body of [professional] men skilled in that particular art)".

4.16 Third, it is often said that where expert opinion fails to provide evidence of a commonly adopted standard or a body of opinion (perhaps because the expert simply gives evidence as to what it would have done) an exception to the *Bolam* test arises. Again, on analysis, this is not a true exception. Rather it describes a situation where one party, usually the defendant, has failed to provide evidence of sufficient probative value to inform the Court as to the standards or practices of the profession. This matter is discussed further below.

4.17 There are very few reported cases involving construction professionals where the courts have found it unnecessary to apply the *Bolam* test. In *Nye Saunders & Partners v Alan E Bristow*,[16] architects were retained to prepare a planning application for the proposed renovation of a house. The employer told the architects that he had about £250,000 to spend and asked for an estimate of the likely cost. The architects consulted a quantity surveyor and gave an estimate of £238,000 but no quantification in respect of future inflation

---

15 The best-known example occurred in the field of solicitors' negligence. In *Edward Wong Finance Co Ltd v Johnson Stokes & Master* [1984] AC 296 (PC), the plaintiff was defrauded by a crooked solicitor. The fraud was possible because of the universally adopted practice in Hong Kong conveyancing of paying over monies before adequate security was obtained. The Privy Council upheld a finding that the defendant solicitors were negligent even though they had acted in accordance with accepted practice.

16 (1987) 37 BLR 92 (CA).

was made. In March 1974, planning permission having been obtained, the employer engaged the architects to provide the services necessary for the completion of the project. In September 1974, the architects gave an up to date estimate amounting to £440,000. The employer then terminated their employment and the architects sued for their fees. The employer contended that the architects had failed in their duty to take care to provide a reliable approximate estimate and in particular to draw attention to the fact that inflation would drive the figure beyond the amount available. The architects' expert gave evidence that this was outside the practice of architects, but the Judge rejected that evidence and found for the employer. The architects appealed. The appeal was rejected. Stephen Browne LJ said:

> [The judge] was entitled to take the view that the evidence of [the architect's experts] did not constitute evidence of a responsible body of architects accepting as a proper practice that no warning of inflation need be given when providing an estimate of the cost of proposed works. It seems to me that the learned judge had ample evidence before him which entitled him to find that there was a failure on the part of Mr Nye to draw the attention of the client to the fact that inflation was a factor which should be taken into account when considering the ultimate cost and that the failure constituted a breach of the Hedley Byrne type duty to the defendant.

4.18 In *Michael Hyde & Associates Ltd v J D Williams & Co Ltd*,[17] JD Williams employed Michael Hyde as architect for the conversion works of cotton mills to storage facilities. These works included the provision of a new direct-fired heating system, which operated by venting the exhaust from gas heaters directly into the building. Unknown other than to a few specialists, this method of heating could result in phenolic yellowing, which in fact occurred and caused discolouration of the goods stored in the mills. British Gas's quotation for the heating system included a disclaimer in respect of discolouration of materials. JD Williams followed this up with British Gas and was reassured that direct-fired heating systems had been used elsewhere without causing discolouration. It discussed the issue with Michael Hyde and they jointly decided to proceed to order and install the direct-fired heating system. As a result of its operation, goods were damaged. JD Williams alleged that Michael Hyde ought to have been aware of the risk of discolouration as a result of the disclaimer and ought to have warned of the risk.

4.19 The judge found that both JD Williams and Michael Hyde had the same level of knowledge – neither knew of phenolic yellowing but both knew that discolouration of materials had occurred in the past, and both believed that British Gas's representative could speak authoritatively and were reassured by him that the risk was not significant. The parties' experts held opposing views on whether Michael Hyde's duty was to draw the risk to JD William's attention, or to investigate further. Michael Hyde argued that these represented two respectable schools of thought, and that since one of them supported its own view it had not been negligent under the *Bolam* test. The Judge, however, considered that the question of breach did not depend on any peculiarities of architectural practice, and that the evidence of Michael Hyde's expert was therefore irrelevant. He held that Michael Hyde had been in breach of duty in failing to investigate the matter of discoloration further.

4.20 The Court of Appeal agreed with this part of the Judge's analysis. It said that there were three situations where the *Bolam*, test did not apply: (a) where expert opinion supporting the practice was not capable of logical support, (b) where the evidence did not constitute evidence of a school of thought, and (c) where the act or omission in question

---

17 [2001] PNLR 8 (CA).

did not require the exercise of any special skill. The Judge had been correct in considering that the case fell into the third category.[18]

SECTION C: ESTABLISHING THE STANDARD

4.21 Evidence of the general practice of the profession is provided to the court or the arbitrator by an "expert"; a construction professional with sufficient experience and knowledge of practice in that field to assist the tribunal to understand the technical issues and make an assessment of the standards and practices adopted by the relevant profession.

4.22 The role and duties of an "expert" are considered in Chapter 8, but it is convenient to note here the particular function of the expert and the limitations of expert evidence.

4.23 First and most important, the tribunal does not delegate to the expert the role of deciding whether a particular act or omission was negligent. The ultimate question is not whether the construction professional's conduct accorded with the practice of the profession. That is merely evidence which assists the court or arbitrator in deciding whether the defendant acted with reasonable skill and care. It is no part of the expert's function to intrude on matters which are the province of the tribunal.

4.24 Secondly it is relatively easy (and often done) for an expert to mistake its obligation to provide evidence as to the practices and standards of the profession with evidence as to what it would have done itself when placed in the circumstances of the defendant. Oliver J said in *Midland Bank Trust Co Ltd v Hett Stubbs & Kemp (a firm)*:[19]

> Clearly, if there is some practice in a profession, some accepted standard of conduct which is laid down by a professional institute or sanctioned by common usage, evidence of that can and ought to be received. But evidence which really amounts to no more than an expression of opinion by a particular practitioner of what he thinks that he would have done had he been placed, hypothetically and without the benefit of hindsight, in the position of the defendant, is of little assistance to the court.

4.25 Those remarks were made in the context of solicitors' negligence, where the court rarely admits expert evidence, but they have resonance in cases involving construction professionals.

4.26 Third expert evidence is only admissible if and to the extent that the expert is qualified to speak to standards and practice in that profession. In *Sansom v Metcalfe Hambleton & Co*[20] Butler-Sloss LJ said:

> a court should be slow to find a professionally qualified man guilty of a breach of his duty of skill and care towards a client (or third party), without evidence from those within the same profession as to the standard expected on the facts of the case and the failure of the professionally qualified man to measure up to that standard. It is not an absolute rule ... but, [un]less it is an obvious case, in the absence of the relevant expert evidence the claim will not be proved.[21]

---

18 The Appeal succeeded on causation. There was insufficient evidence to justify the Judge's finding that had Michael Hyde investigated it would have discovered the risk of phenolic yellowing.
19 [1979] Ch 384 at 402.
20 [1998] PNLR 542 at 549.
21 Similar observations were made in *Investors in Industry Commercial Properties Ltd v South Bedfordshire District Council* [1986] QB 1034 (CA), where evidence given by expert engineers was held to have little weight in a case against architects.

4.27 Each case will be taken on its facts but, as a general rule, the more remote the "expert" from the field of practice which is the subject of the claim, the less probative its evidence.[22]

4.28 Notwithstanding those limitations the courts regard the obtaining of expert evidence as a necessary precursor of bringing a claim against a construction professional. In *Pantelli Associates Ltd v Corporate City Developments Number Two Ltd*,[23] Coulson J indicated that commencing proceedings in these circumstances would be an abuse of process. It is respectfully suggested that this goes too far[24] and subsequent decisions have somewhat softened the court's approach.[25] However, any potential claimant would be most unwise to commence proceedings against a construction professional without the clear support of expert evidence and (absent very unusual circumstances) it would be even more unwise to persist in the claim without obtaining that evidence.

4.29 The *Pantelli* case (and other cases involving experts)[26] illustrate the important point that the courts control the use of expert evidence in the course of proceedings just as they decide its utility in the final result. The Technology and Construction Court ("TCC"), as the specialist court for construction cases, is staffed by Judges very experienced in construction related disputes who are perhaps more ready to draw on their own expertise when considering expert evidence than might be Judges with less exposure to that world. As a general guide to the attitude of the Court, it is suggested that a useful summary of the role of expert evidence in TCC cases concerned with the liability of construction professionals is provided by His Honour Judge Humphrey Lloyd QC in *Royal Brompton Hospital NHS Trust v Hammond and ors (No. 9)*:[27]

> First, the *Bolam* test applies where the court cannot answer the question without expert evidence as to the body of professional practice prevailing at the time where the negligence lies in not following established practice. It is required both to prove that practice, as a matter of evidence, since the court would not otherwise know of it either as a matter of common sense (or judicial notice) or as a matter of expertise which the court should possess ... Secondly, it is needed in certain situations, since as a matter of policy, since a professional person should not be held liable without the court being satisfied that any competent professional would have done otherwise and that as a result the consequences of the negligence would not have occurred. However ... it cannot be an absolute requirement without which the court cannot reach its decision. Thirdly, however, expert evidence may be needed to help the court assess the available evidence, such as, in a case of professional negligence, by indicating what factors or technical considerations would influence the judgment of a professional person, or, in other instances, aspects of the way in which construction work is executed which might affect findings of fact, e.g. as to the extent of delay or disruption. That evidence is not needed at all where the decision is a matter of common sense ... but it is helpful where the allegation does not require or should not depend on evidence of established practice, such as, in a case of

---

22 Note the approach taken by Ramsey J to the evidence of a specialist geotechnical engineer in a case involving negligence by a structural and civil engineer in *Cooperative Group Ltd v John Allen Associates Ltd* [2010] EWHC 2300 (TCC) at [142]–[149].

23 [2010] EWHC 3189 (TCC).

24 Pursuit of any claim which the claimant knows is doomed to fail is an abuse of process. Similarly it is an abuse of process to raise a claim which unintelligible so that the defendant does not know the case it has to meet. Either of these states of affairs may be present when professional negligence proceedings are commenced without the benefit of expert evidence, but they are not bound to be.

25 See *ACD (Landscape Architects) Ltd v Overall* [2012] EWHC 100 (TCC) and *Whessoe Oil and Gas Ltd v Dale* [2012] EWHC 1788 (TCC).

26 See Chapter 8.

27 [2002] EWHC 2037 (TCC) at [20].

professional negligence, where there is no such practice and therefore the court has to understand what would go through the mind of a professional person in those cases where what would be common sense to the rest of the world would not or might be not sensible in that profession or occupation. The factors or considerations that would need to be borne in mind might therefore be rehearsed for completeness. It may also be desirable to present that evidence so that the judge knows that his conclusion is not inconsistent with the views of others whose opinions deserve respect. However, in those cases, whilst most claimants will provide expert evidence, it is not indispensable and a party may succeed without it ... The court is able to form its own view, and is entitled to do so, without the need for such evidence of practice or opinion from an expert for this is the territory where the purposes or the upshot of the opinion is no more than a statement of belief as to he or she would have done in the circumstances, presented as evidence of practice. Thus judges of this court possess and acquire knowledge of the construction industry and other areas of commerce so that they do not require such expert evidence to enable them to assess evidence, not only in relation to common questions of professional practice but also in relation to matters such as causative impact of events in cases of delay or disruption or ordinary practice in the construction and other fields of commercial disputes which come to this court. If that were not the case there would be little point in having specialist courts, such as the TCC.

SECTION D: CODES OF PRACTICE

4.30 Plainly, because the court or arbitrator is concerned with the standards and practice of the profession industry, codes of practice, British Standards and regulatory requirements will be of the utmost importance. The construction industry is unusual in the large number and range of codes and standards which are intended to provide information to construction professionals and to guide them as to issues of safety and good practice. It is very common for allegations of breach of duty against construction professionals to take the form of failing to apply or adequately to note the contents of particular codes or standards.

4.31 If satisfied that a code or standard was applicable to a particular set of facts, but was departed from or overlooked by the construction professional, courts and arbitrators will almost always find a failure to exercise reasonable skill and care. The practical effect of the failure to comply with any relevant code or standard is to shift the burden of proof to the construction professional to show that notwithstanding the failure it acted with reasonable care and skill. In the New Zealand case of *Bevan Investments Ltd v Blackhall and Struthers (No2)*,[28] the court was concerned with allegations of negligent design by engineers and it was contended that the engineers had failed to properly to apply certain codes. Beattie J said:

> Bearing in mind the function of codes, a design which departs substantially from them is prima facie a faulty design, unless it can be demonstrated that it conforms to accepted engineering practice by rational analysis. If I am correct in this appreciation, and if on the evidence it is established that the design in several material respects fails to comply with the relevant codes, then [the defendant engineer] and his experts must show that the design is capable of rational analysis and is adequate and safe. I do not go so far as to say however that the mere circumstance that a client is unaware that the designer is working outside a code can of itself categorise a designer's actions as negligent.[29]

4.32 Of course, the fact that a failure to comply with a code or standard is evidence of a failure to exercise reasonable skill and care does not mean that compliance is always

---

28 [1973] 2 NZLR 45.
29 Ibid. at 65–66.

a satisfactory defence to an allegation of breach of duty. In *Holland Hannen & Cubitts (Northern) Ltd v Welsh Health Technical Services Organisation*[30] Robert Goff LJ said:[31]

> The structural engineer will therefore simply consider the profile of the floor as such; and ask himself the question whether there is a significant risk that the floor, with that profile, in the building in question, may be unacceptable. In considering that question he cannot simply rely on the codes of practice. It is plain from the evidence that the code of practice is no more than a guide for use by professional men, who have to exercise their own expertise; this must moreover be especially true in a case such as the present, where the design was a novel one, omitting as it did a finishing screed. Practice alone can, I consider, provide of itself no reliable guide where, as here, a novel design concept is being used.

4.33 One of the key functions of an expert witness will often be to explain the meaning and application of a code or standard. Despite the best efforts of the persons who draft these documents, they are sometimes unclear or even contradictory. Unlike contracts or legislation these are not legal documents which the court or arbitrator construes by reference to established principles. Experts from the relevant field will often be required to give evidence as to what the industry generally understands a particular part of a code to mean and/or how in practice one part of a code or standard impacts upon another part. That position is to be contrasted with statutory codes (particularly prevalent in fields such as health and safety and environmental regulation). These legislative provisions are matters of law. They have a particular meaning which the court or arbitral tribunal ascertains by applying legal principles of construction. That process may lead to a meaning which has not been understood or followed by practitioners in the field.

SECTION E: THE BAND OF REASONABLE ACTIONS

4.34 In the paradigm professional negligence claim there are two courses of action open to the construction professional and the only question is whether the course of action taken was competent. In the real world, however, that is not always the case. In many cases there are a *range* of ways in which a construction professional might competently deal with a particular issue.

4.35 An engineer preparing foundation design may have a choice of design solutions, three of which would be solutions adopted by a responsible body of the profession and one of which would outside that standard so that adopting it would be a breach of duty. An architect may be required to undertake occasional inspections of work being undertaken but reasonably have a discretion as to when and what it inspects.

4.36 This matters because of causation. In a professional negligence claim against a construction professional it is rarely sufficient for the claimant merely to contend that the advice or action taken by the construction professional was negligent. It will usually be necessary for the claimant to contend as to what the construction professional should have done instead. It is on that basis that the claimant can establish the loss it has suffered (because damages are assessed by putting it in the position it would have been in had the construction professional acted competently).

---

30 (1986) 35 BLR 1 (CA).
31 Ibid. at 25.

4.37 It follows that the application of the *Bolam* test to any particular case may be a broader one than first appears: in addition to asking whether the act or omission was consistent with the standard of care demonstrated by the ordinarily hypothetical member of the profession the tribunal may need to go further and ask what alternatives were available to that ordinarily hypothetical member of the profession.

SECTION F: OUTCOME, NOT METHODOLOGY

4.38 Cases involving allegations of negligence against construction professionals frequently involve a minute analysis of precisely how the construction professional came to the design or action which is the subject matter of the case. The context and documentation leading up to the event are scrutinised and the construction professional is closely cross-examined as to the thought processes involved in reaching the event.

4.39 It may be somewhat surprising, therefore, that the *Bolam* test is not concerned with how the construction professional reached the decision it did. It is solely concerned with whether that decision was one which an ordinarily competent construction professional facing the same set of circumstances could have reached.

4.40 The focus of the Court is fixed upon the act or omission of the construction professional which caused the loss, not the loss itself or the steps taken prior to the act, or omission. As to the former, just as is other fields of professional negligence, there is no rule of law or practice that just because something has gone wrong the professional adviser must have been negligent.[32] Lord Salmon said in *Sutcliffe v Thackrah*:[33]

> It by no means follows that a professional valuation was negligently given because it turns out to have been wholly wrong. Nor does the fact that an architect's certificate was given to the wrong amount of itself prove negligence against the architect.

As to the latter, a design is competent if it is one which could be produced by the ordinarily competent designer and it does not matter if the competent design has been produced by a series of incompetent acts.

4.41 This aspect of the *Bolam* test is best illustrated by an example. An engineer is asked to design a floor slab which will bridge between supports rather than be consistently supported. The engineer is tired and under time pressure. Rather than carry out calculations, it makes a guess at the various loading and stress parameters. The slab fails and the employer sues. The engineer admits that the way in which it carried out its task was not in accordance with engineering practice, but provides clear expert evidence that, by pure luck, the guessed calculations corresponded closely to the results which would have been achieved had the exercise been carried out carefully and using procedures widely adopted at the time. Because of advances in knowledge as to the behaviour of suspended slabs of this kind subsequent to the design it has become apparent that those procedures were inadequate,

---

[32] This approach is sometimes referred to by the latin tag of "res ipsa loquitur": the thing speaks for itself. It never does. The legal burden of proving negligence and causation always lies with the claimant. For an example of a case where the claimant failed to persuade the court that the circumstances must indicate negligence see *Copthorne Hotel (Newcastle) Ltd v Arup Associates (No. 1)* (1996) 58 Con LR 105.

[33] [1974] AC 727 (HL) at 760.

but that information was not available at the time. On an application of the *Bolam* test, the engineer was not negligent.

4.42 In *Adams v Rhymney Valley District Council*,[34] the claimants were tenants of the defendant authority. Shortly before their tenancy commenced, the defendants had replaced the windows in the claimants' property in common with most of the windows in their housing stock. The design of window adopted was one that involved a window lock with a removable key, rather than a lock with a push-button release on the inside. No smoke alarms were fitted in the building. In October 1991, in the course of a fire at the property, the first claimant father, who was downstairs, was able to escape; he was unable to go upstairs with a key for the windows; the second claimant mother (who had been asleep upstairs with the three children) eventually managed to break one of the windows, but fell out, suffering severe injury. The couple's three children died in the fire. The claimants brought an action, alleging negligence by the authority in not taking expert opinion, and in not fitting windows with push-button release locks.

4.43 At first instance, the judge accepted the expert evidence called on behalf of the authority to the effect that the type of window lock adopted was one a competent designer might have adopted in 1989/90, and that in the circumstances, the local authority could not be said to have been negligent. The claimants appealed. They argued that the local authority was negligent because it never considered the relevant British Standard or thought about push-button release locks.

4.44 The Court of Appeal by a majority rejected that argument. In order to avoid a finding of negligence, a professional person need not necessarily review the various bodies of responsible opinion; it was sufficient that he adopted a course supported by a body of responsible professional opinion.[35] Sir Christopher Staughton noted that it made no difference that the defendant lacked the special skill under consideration and,

> Nor can it be a requirement of the Bolam test that the defendant should have considered and reflected upon the alternative courses available and made a conscious choice between them. Seeing that, upon the hypothesis which is inherent in the problem, a respectable body of professionals have been and are in favour of each course, I do not see that the defendant is required to go through the same thought process in order to deserve the support of those who favour the course which he chooses. The consultant who naturally chooses one method out of long experience is as much entitled to rely on the Bolam doctrine as one who sits down in his chair and goes through the whole process of choice again...
>
> The efforts of the council's team produced a design which was not a negligent design. It was one which skilled and careful people had produced in the majority of cases up and down the country. In those circumstances there was in my judgment no negligence in their failure to consult with others or in the other matters relief on. An ordinary builder who is asked to make a window of new design, or to insert window locks recommended by the local crime prevention officer, is not negligent if he produces a product which is acceptable by the standards of the day. I do not see that reasonable skill and care means something different for the council's Technical Services Department.[36]

4.45 It will be only in limited number of exceptional cases that the law as contained in *Adams v Rhymney* is relevant to a case concerning construction professionals. Where it is

---

34 [2001] PNLR 4 (CA).
35 Sedley LJ dissented stating that because the Council had not exercised any special skill and care the *Bolam* test did not apply.
36 Paragraphs 43 and 50.

plain that the construction professional either overlooked a relevant matter, or went about its task in the wrong way, and it is plain that the loss or damage was causally related to the construction professional's act or omission (in the sense that a different act or omission would have led to a different result), the construction professional may find that it bears the evidential burden of establishing that its design was not negligent. The reason that the process by which the construction professional came to its design is the subject of minute examination is that errors in the process may be evidence from which the court can infer that the result fell below the standard of reasonable skill and care. Absent evidence to rebut that inference, the court is likely to find that the result was negligent because it was negligently arrived at.

SECTION G: SENIORITY, RESOURCES AND PRICE

4.46 On its face, the *Bolam* test applies a uniform standard: the hypothetical competent architect, engineer or project manager is expected to perform to a uniform standard notwithstanding that the defendant professional may be vastly experienced, exhaustively resourced and extremely well paid, or by contrast, nearly a novice, acting as a one-man band and working for a pittance.

4.47 This carries a degree of artificiality. If I order a meal at a Michelin-starred restaurant I am entitled to expect a higher quality of service than if I order in the local café. In the same way (it might be thought) I could expect a higher standard of care if I engage an architect from a London firm with a worldwide reputation than if I engage the newly qualified individual practitioner who lives down the road.

4.48 This artificiality, it is suggested, is partly rooted in the implicit assumption that, by reason of its membership of a profession, a professional is bound to adhere to particular standards and to adopt particular procedures which enable the court to establish a minimum threshold common to all. It is more obviously rooted in the difficulties which would attach to measuring the acts and omissions of different professionals within the same profession to different standards by reason of their experience, sophistication or position in the market.

4.49 The courts in England and Wales have been astute to maintain at least the theory of a common standard whilst, in practice, allowing a degree of flexibility. Thus in other fields of professional negligence there is an unresolved tension between cases which rigidly adhere to the line that the professional's actual qualifications and experience are irrelevant[37] and cases which recognise the economic reality that different professionals in the same fields offer (and are expected to provide) different quality of service.[38]

---

37 See, for example, *Andrew Master Hones Ltd v Cruikshank & Fairweather* [1980] RPC 16. On appeal, the defendant conceded that it had been negligent.

38 In *Matrix Securities Ltd v Theodore Goddard* [1998] PNLR 290 Lloyd J held that a firm of city solicitors giving tax advice were not obliged to work to a higher standard than the reasonable skill and care of the ordinarily competent solicitor but that some account should be taken of their expertise: they should be judged to the standard of an ordinarily competent firm of solicitors with a specialist tax department. Subsequent cases have, in effect, required specialist firms of solicitors to work to specialist standards of care: see *Agouman v Leigh Day (A Firm)* [2016] EWHC 1324.

4.50 In the field of construction professionals the same position pertains: the standard to be achieved is the same whatever the experience and sophistication of the construction professional.[39] However, economic reality is likely to be reflected in two ways.

4.51 First, many professional engagements qualify the standard of care to be provided by using wording such as "the reasonable skill and care of an architect experienced in projects of this kind". That allows the employer to assert that the larger and more sophisticated the project the more "experience" (used here as a synonym for "care") it was entitled to expect from its construction professional.

4.52 Courts are likely to follow this trend even where such wording is absent: if a construction professional holds itself out to carry out work in the context of a particularly complex and expensive project it should possess the skills to the level commonly expected of construction professionals carrying out that kind of work.

4.53 Secondly, the fact that a construction professional has ready access to specialist expertise may serve to fix that professional with that expertise and make it more difficult for that professional to argue that the exercise reasonable skill and care should not take account of that attribute. Thus, a firm of consulting engineers which possesses geotechnical expertise may in practice be held to a higher standard of care in respect of geotechnical issues than an engineer which does not even though the terms of their engagements may be identical.

4.54 It is relatively unlikely that the courts will ever sanction a lower standard of care because, for example, the construction professional was, to the knowledge of the client, inexperienced and engaged by the client because the client did not want to pay for someone of greater experience. Absent some very clear provisions in the contract of the construction professional the courts are likely strictly to enforce the line that any construction professional which practices in an area where it lacked the experience or expertise ordinarily required to carry out that work has no business carrying out that practice and will be held to the standard required to effect it.

SECTION H: DELEGATION

4.55 The complexity of modern construction projects means that it is usual for a substantial number of persons with different expertise to have an input into design and construction decisions. In many cases the construction professional's contract with its employer will clearly delineate the responsibility of the construction professional where some aspect of the works is to be designed by others. However, it is very common for architects or engineers to be contractually responsible for designs or other decisions which require input from specialist advisers.

4.56 The general rule, in such circumstances, is that the construction professional is entitled to rely upon the advice of the specialist. In *Investors in Industry Commercial Properties Ltd v South Bedfordshire District Council*,[40] one of the issues was the extent to which the architects were entitled to rely upon the advice of engineers.[41] Slade LJ said:

> Whether or not in any given instance [the RIBA Conditions] apply, it must generally be the duty of an architect to exercise reasonable care in the work which he is engaged to perform.

---

39 See His Honour Judge Newey QC in *Board of Governors of the Hospitals for Sick Children v McLaughlin & Harvey plc and ors* (1987) 19 Con LR 25 at 92.
40 [1986] QB 1034 (CA).
41 The engineers, who were the natural defendants, were uninsured.

However, Clauses 1.20, 1.22 and 1.23 of the RIBA conditions, in our judgment, clearly contemplate that where a particular part of the work involved in a building contract involves specialist knowledge or skill beyond that which an architect of ordinary competence may reasonably be expected to possess, the architect is at liberty to recommend to his client that a reputable independent consultant who appears to have the relevant specialist knowledge or skill shall be appointed by the client to perform this task. If following such a recommendation a consultant with these qualifications is appointed, the architect will normally carry no legal responsibility for the work to be done by the expert which is beyond the capability of an architect of ordinary competence; in relation to the work allotted to the expert, the architect's legal responsibility will normally be confined to directing and coordinating the expert's work in the whole. However, this is subject to one important qualification. If any danger or problem arises in connection with the work allotted to the expert of which an architect of ordinary competence reasonably ought to be aware and reasonably could be expected to warn the client despite the employment of the expert, and despite what the expert says or does about it, it is in our judgment, the duty of the architect to warn the client. In such a contingency he is not entitled to rely blindly on the expert, with no mind of his own, on matters which must or should have been apparent to him.[42]

4.57 It is suggested that this statement of principle is not confined to architects engaged under RIBA conditions and indeed is of general application to construction professionals. However, the statement is necessarily subject to two qualifications.

4.58 The first and most important comment is that this aspect of the *Bolam* test is not confined in its application to circumstances where the construction professional relies upon specialist advice. It is equally important (possibly more important) in situations where the construction professional fails to advise that specialist advice be obtained. Underlying both applications is the question of whether, in any particular circumstance, the ordinarily competent construction professional would necessarily recognise that it lacked the expertise adequately to carry out its function without the assistance of expert advice. This is always a question of fact, decided by the court with the assistance of expert evidence.[43]

4.59 The second qualification is that whether delegation is competent may be a subsequent question, the prior question being whether it is *permitted*. The construction professional's contract may provide the answer, but it most cases it will be silent. If the construction professional has obtained the client's agreement (and has done so in a manner which permits the client to understand its position) no issue arises. But in a substantial body of instances the construction professional will have neither clear guidance from the contract, nor the permission of its client.

4.60 This matters, of course, because the client's contract is with the construction professional, not the specialist adviser. If the advice of the specialist is wrong, the client may have no remedy. There is an unavoidable tension between the client's desire for a competent

---

42 At 807.
43 For example, in *Richard Roberts Holdings Ltd v Douglas Smith Stimson Partnership & Others* (1988) 46 BLR 50 architects were liable for the defective lining of a tank because they had approved the recommendation of the specialist supplier without making proper enquiries. They should either have warned the employer that they lacked the expertise to assess the suitability of the proposed lining and should have advised that specialist advice be obtained from a chemist. In *Try Build Ltd v Invicta Leisure (Tennis) Ltd* (1997) 71 Con LR 141 engineers failed to appreciate that novel features of a design for a sports hall roof developed by a specialist supplier had not been successfully developed and were inadequate. The engineers were found to have been negligent in failing to check and warn of deficiencies in the designs prepared by specialist sub-contractors.

design or service and the its desire to have a single person who is legally responsible for providing that design or service.

4.61 The authorities have come down on different sides of this equation, often on the basis of subtle factual differences. In *Moresk Cleaners Ltd v Hicks*[44] the architect delegated the design of the reinforced concrete frame of the building to a structural engineer. The design was negligent and the architect was held liable. By contrast in *Merton LBC v Lowe*[45] the architect's design included suspended ceilings surfaced with a proprietary product which had been recommended by a specialist sub-contractor. This product proved unsuitable but the employer's action against the architect failed. It had acted reasonably in accepting the recommendation of the specialist and, unlike in *Moresk*, had not delegated "the whole task of design" to another.

4.62 In *Cooperative Group Ltd v John Allen Associates Ltd*,[46] Ramsey J. reviewed the authorities referred to above, as well as some Canadian decisions, and derived the following propositions:

- a construction professional does not by the mere act of obtaining advice or a design from another party thereby divest himself of his duty in respect of that advice or design;
- a construction professional can discharge his duty to take reasonable care by relying upon the advice or design of a specialist provided that he acts reasonably in doing so;
- in determining whether a construction professional acted reasonably in seeking the assistance of specialists to discharge his duty to the client, the court has to consider all the circumstances, including (a) whether the assistance was taken from an appropriate specialist, (b) whether it was reasonable to seek assistance from other professionals, research or other associations or other sources, (c) whether there was information (relating to the assistance) which should have led the professional to give a warning, (d) whether and to what extent the client might have a remedy in respect of the advice from the specialist consulted, and (e) whether the construction professional should have advised the client to seek advice elsewhere or should himself have taken professional advice under a separate retainer.

4.63 It is suggested that this is the approach which will be followed by courts and arbitrators in all cases where permission to rely upon the advice of specialists is either not expressly provided by the contract[47] or obtained from the client.

SECTION I: DUTY TO REVIEW

4.64 For obvious reasons the *Bolam* test focuses upon the point in time when the construction professional's design is finalised or its action or omission takes place. However, it

---

44 [1966] 2 Lloyd's Rep 338.
45 (1982) 18 BLR 130 (CA).
46 [2010] EWHC 2300 (TCC).
47 Or expressly prevented – although this is unlikely.

is not immediately apparent why this should be the only point in time at which the design, action or omission falls to be assessed.

4.65 Construction projects have a long gestation. A design prepared in year 1 of the project may not be constructed until year 4. In that period new information may come to light, both as to the efficacy of the technical underpinning of the design, action or omission and as to the effect of other factors (for example, evidence that a complementary design prepared by some other construction professional is not likely to perform as predicted). In any event, if a construction professional remains engaged in a project does it not have a continuing duty to keep its design under review?

4.66 These questions have generally arisen in the context of disputes where the date of breach matters for limitation purposes. In other fields of professional negligence there has been debate as to whether a duty to exercise reasonable skill and care exists as a duty to review an action which has been taken and if so whether it continues until the end of the retainer. In *New Islington and Hackney Housing Association Ltd v Pollard Thomas and Edwards Ltd*,[48] Dyson J had to consider the extent of an architect's continuing duty to review his design:

> I accept the proposition that, although it is necessary to look at the circumstances of each engagement, a designer who also supervises or inspects work will generally be obliged to review that design up until that design has been included in the work: see Jackson and Powell on Professional Negligence 4th Edition para. 2.17. In a number of cases, it has been held that this duty continues until practical completion…
>
> But it is necessary to consider the scope of that duty in a little more detail. What does the duty to review the design entail? In what circumstances will an architect be in breach of that duty? I find it convenient to consider an example. Let us suppose that an architect is engaged on the standard RIBA Conditions of Engagement to provide the full service (as PTE were in the present case), including administering a building contract in a standard JCT form of contract. Suppose that he designs the foundations of a building (a large office block), the foundations are constructed in accordance with his design, and several years later, practical completion is achieved. Let us further suppose that the design of the foundations is defective and one which no reasonably competent architect would have produced: in other words, the architect was negligent. There can be no doubt that the architect commits a breach of contract when he completes the design and gives instructions to the contractor to construct the foundations in accordance with it. But in what sense and to what extent is the architect under a duty to review his negligent design once the foundations have been designed and constructed?
>
> In my view, in the absence of an express term or express instructions, he is not under a duty specifically to review the design of the foundations, unless something occurs to make it necessary, or at least prudent, for a reasonably competent architect to do so. For example, a specific duty might arise if, before completion, the inadequacy of the foundations causes the building to show signs of distress; or if the architect reads an article which shows that the materials that he has specified for the foundations are not fit for their purpose; or if he learns from some other source that the design is dangerous. In such circumstances, I am in no doubt that the architect would be under a duty to review the design, and, if necessary, issue variation instructions to the contractor to remedy the problem. But in the absence of some reason such as this, I do not think that an architect who has designed and supervised the construction of foundations is thereafter under an obligation to review his design.

---

48 [2001] BLR 74 (TCC).

I do not accept that in every case where an architect has negligently introduced a defective design into a building, he is also by the same token in breach of a continuing breach of a contractual obligation to review his design. In *Midland Bank v Hett Stubbs & Kemp*,[49] 403C, Oliver J said:

> "It is not seriously arguable that a solicitor who or whose firm has acted negligently comes under a continuing duty to take care to remind himself of the negligence of which, ex hypothesi, he is unaware".

In my view, that observation is as apt to apply to an architect as it is to a solicitor. The position is quite different where the architect (or solicitor) knows, or ought to know, of his earlier negligence. When that occurs, then he may well be under a contractual obligation to review his earlier performance, and advise his client honestly and competently of his opinion. Whether he is in fact under such a duty when he has actual or constructive knowledge of his earlier breach of contract will depend on whether the contract is still being performed. If the contract has been discharged (for whatever reason), then the professional person may be under a duty in tort to advise his client of his earlier breach of contract, but it is difficult to see how he can be under any contractual duty to do so…

In my judgment, the duty does not require the architect to review any particular aspect of the design that he has already completed unless he has good reason for so doing. What is a good reason must be determined objectively, and the standard is set by reference to what a reasonably competent architect would do in the circumstances.[50]

4.67 The position is therefore as follows:

- the duty to review design will arise only if the designer has good reason to reconsider his original design;
- good reason includes the situation where something happens to call into question the efficacy of the design which comes to the designer's attention;
- good reason will also include the situation where the designer ought to know that there may be something wrong with the design because of its carelessness (but this constructive knowledge will not arise merely because it has been careless);
- "good reason" is a question to be addressed by the application of the *Bolam* test;
- in any event, the duty ceases once the designer ceases to be engaged.

SECTION J: SKILL AND CARE NOT ENOUGH

4.68 As has been stated above, the contracts of construction professionals almost invariably require the carrying out of services to the standard of reasonable skill and care. However, whilst this is true as a statement of fact for written contracts and as a statement of law for contracts which contain no express terms governing the standard by which the construction professional is to be judged, it is none the less possible for the employer to require performance of a higher standard. Indeed it is possible for the employer to require performance of an absolute standard.

---

49 [1979] Ch 384 at 403C.
50 Dyson J considered the statement of Sachs LJ in *Brickfield Properties v Newton* [1971] 1 WLR 862 at 973 that:

> The architect is under a continuing duty to check that his design will work in practice and to correct any errors which may emerge. It savours of the ridiculous for the architect to be able to say, as it was here suggested that he could say: "true, my design was faulty, but, of course, I saw to it that the contractors followed it faithfully" and be enabled on that ground to succeed in the action.

As he pointed out, in that case the Court was not concerned to explore the scope of an architect's continuing duty to review his design and so the statement cannot be used to support a continuing duty of retrospective review.

4.69 In *Platform Funding Ltd v Bank of Scotland plc*,[51] the issue was whether the valuer was required to exercise reasonable skill and care to value a particular property or to value a particular property and to do so with reasonable skill and care. The Court of Appeal held it was the latter. Moore Bick LJ observed:[52]

> It has been observed on many occasions that those who provide professional services do not generally give an unqualified undertaking to produce the desired result. Thus in *Greaves & Co (Contractors) Ltd v Baynham Meikle & Partners*[53] Lord Denning MR said, at p. 1100:
>
>> "Apply this to the employment of a professional man. The law does not usually imply a warranty that he will achieve the desired result, but only a term that he will use reasonable care and skill. The surgeon does not warrant that he will cure the patient. Nor does the solicitor warrant that he will win the case".
>
> I am inclined to think that the reason why the law does not ordinarily construe the contract in such cases as giving rise to an unqualified obligation owes more to the nature of the services themselves, the context in which they are to be provided and the fact that the desired result is not one which any professional person can reasonably guarantee, than to the fact that the provision of the services involves the exercise of special skill. In other contexts the law has no difficulty in implying an unqualified obligation to achieve the desired result. For example, if one employs a skilled craftsman to make a table, the law will normally imply a term that the table will be reasonably fit for its purpose; and as Lord Denning MR observed, again in *Greaves & Co (Contractors) Ltd v Baynham Meikle & Partners*: "when a dentist agrees to make a set of false teeth for a patient, there is an implied warranty that they will fit his gums: see *Samuels v Davis*.[54]
>
> In principle, therefore, although there is every reason to assume, in the absence of a term to the contrary, that a professional person has undertaken no more than to use reasonable skill and care in relation to matters calling for the exercise of his professional skill and expertise, there would seem to be no good reason why one should make a similar assumption in relation to other aspects of his instructions. As to those, one would expect to construe the terms of engagement in each case to ascertain the precise nature of the obligations undertaken, without making any prior assumption that they are qualified or unqualified. The engagement of a photographer to take a portrait photograph of a particular person provides an illustration. The photographer undertakes no more than to exercise reasonable professional skill and care in and about the creation of the image, but there is no obvious reason why one should assume, in the absence of a term to the contrary, that he did not accept an unqualified obligation to photograph the right person. On the contrary, one might think it more natural that he should have done so.

4.70 Thus the failure by the construction professional to carry out part of a service – for example, an inspection – may not be a breach of its obligation to exercise reasonable skill and care but a breach of its obligation to carry out certain work.[55]

4.71 Moreover it is possible that the construction professional's obligations will be construed in the context of its client's obligations where the contract professional is aware of these. That may impose higher standards of care. In *Consultant's Group International v*

---

51 [2009] QB 426 (CA).
52 At [17] to [19].
53 [1975] 1 WLR 1905 (CA).
54 [1943] KB 526.
55 In *CFW Architects v Cowlin Construction Ltd* [2006] EWHC 6 (TCC), it was held that, on the true construction of the architects' appointment in they had warranted that their design drawings would be provided by a particular time, and not merely that they would exercise reasonable skill and care to meet the target date.

Worman (John) Ltd,[56] an architect was engaged by a contractor to design an abattoir. The contract between the employer and the contractor contained onerous criteria (so that the building might qualify for grant aid). The architect's contract with the contract was construed as containing an obligation to design the building to meet those criteria.

4.72 The particular area of risk to construction professionals lies in contract which define the construction professional's obligations both by reference to reasonable skill and care any by reference to the achievement of some performance standard.

4.73 In *MT Hojgaard v E.ON Climate & Renewables UK Robin Rigg East Ltd* a contractor was engaged to design and install wind turbines and relied in its foundations design upon an international standard. Unknown to the industry, the standard contained a fundamental error. The contract included an express term requiring the contractor to exercise reasonable skill and care in design; but it also included a requirement for foundations with a 20-year service life. The designer's reliance on the standard, albeit consistent with the exercise of due skill and care in design, meant that the foundations did not have the required service life. The Court of Appeal[57] held that, properly construed, the requirement for a 20-year service life was not a guarantee that the foundations would operate for 20 years. Accordingly, the contractor was not liable for the cost of remedial works. The Supreme Court[58] disagreed. On its true construction the contract imposed differing requirements, all of which had to be complied with.

4.74 *Hojgaard* was concerned with the obligations of a design and build contractor. The argument was mainly concerned with the way in which clauses of a commercial contract are read together. It remains to be see whether the same strict approach will be taken with a construction professional who is a pure designer. It is suggested that there is no reason in principle why such the issue should not be one of pure construction, albeit from the starting point that construction professionals generally owe no higher duty than to exercise reasonable care and skill.[59]

---

56 (1985) 9 Con LR 46.
57 [2015] EWCA Civ 407.
58 [2017] UKSC 59.
59 In *Costain Ltd v Charles Haswell & Partners Ltd* [2009] EWHC 3140 (TCC), a civil engineer's written appointment was construed as imposing strict obligations as to the standards to be met by its design. The engineer's appointment contained a warranty that: Any part of the works designed pursuant to this Agreement if constructed in accordance with such design, shall meet the requirements described in the Specification or reasonably to be inferred from the Tender Documents or the Contract or the written requirements of *Costain* and designed in accordance with good up to date engineering practice and with all applicable laws, by laws, codes or mandatory regulations and in all respects with the requirements of the Contract.

CHAPTER 5

Common issues in a construction project

Introduction	79
Section A: site investigation and specialist survey	80
Section B: planning and regulation	82
Section C: budgets and estimates	83
Section D: design	87
Introduction	87
Whose obligation?	87
The standard to be achieved	88
Achieving the client's objectives	88
Specific design attributes	90
Coordination and integration of design	92
Novel design	93
The obligation of the reviewer	94
Inspection or supervision by the designer	95
Approval of defective design	95
Section E: the tender process	96
Introduction	96
Specification	97
Bills of quantities	97
Selection of contractors	98
Construction contracts	98
Other contracts	99
Section F: contract administration	100
Introduction	100
Familiarisation	101
Providing information and giving instructions	102
Certification	103
Inspection	106
Reporting to the client	110
Keeping records	110

INTRODUCTION

5.1 Claims against construction professionals cover an enormous range of circumstances and, because they generally turn upon findings concerning the exercise of reasonable skill and care, they are highly fact sensitive. The outcome of one reported case should be taken as no more than an illustration of how the court might treat another case with broadly similar facts.

5.2 Whilst it might be possible to consider particular types of case thematically, by reference to different construction professionals, it more convenient to examine the courts' treatment of disputes arising at different stages of a construction project. This is particularly useful where the function under discussion (for example, contract administration) might be carried out be any one of a number of construction professionals in different disciplines (in that case, architect, engineer or a project manager).

5.3 A broadly chronological analysis has its disadvantages. Many processes or functions in a construction project run concurrently or near concurrently. There are relatively few clear delineations and design and design coordination, in particular, are iterative. Most construction disputes arise out of a complex factual background which involves the consideration of a number of different facts and matters arising during the project.

SECTION A: SITE INVESTIGATION AND SPECIALIST SURVEY

5.4 One of the earliest tasks on a construction project is the measurement and investigation of the site. This involves (1) physical measurement and survey of the site, its boundaries and any buildings situated within it and (2) investigation and ascertainment of the ground conditions.

5.5 However, the obligations of construction professionals engaged at this early stage may not be concerned with the actual measurement or investigations. They may be concerned with obtaining of such information and the analysis of what it shows. The client may have undertaken a rudimentary survey of the site (probably for valuation purposes) but the architect or project manager will need to decide what kind of detailed survey is required. There may be historic records of the site conditions (particularly if the site has previously been the subject of development) but the project manager or engineer will need to decide precisely what further investigations are required and indeed, having obtained these investigations, to make an assessment of what further inquiries are prompted.

5.6 All of these processes have to be undertaken with an appreciation of the client's ultimate intentions for the site and an understanding of the outline design (such as it is at this stage). Many of them have to be undertaken in conjunction with an ongoing planning process (see below). A surprising number of claims against architects, engineers and project managers have their origins in mistakes and missed opportunities in this early stage.

5.7 It might be thought that an assessment of the physical constraints of the site would be an area unlikely to throw up complexity. However, that would be to underestimate the economic importance of site constraints in construction projects.

5.8 At its simplest level, the physical boundaries of the site may constrain the size and footprint of the building which the client may wish to erect. Planning or other legal restrictions may limit the land available for construction.[1] Adequate site access, which means adequate access to the highway may impose further constraints. The proposed process of construction must be thought through and adjusted to the site. All construction projects have a critical path and the length and vulnerability of the critical path equates to cost. Particularly for larger projects, a well-planned sequence of construction involves a

---

1 In *John Harris Partnership v Groveworld Ltd* [1999] PNLR 697 (TCC), the architects mistakenly drew up plans to include land outside the boundary of the site. This caused a need to revise the design and incur costs.

consideration not just of what can be undertaken on the site, but in what order. The presence of certain physical features of the site may require careful consideration.[2]

5.9 If any part of the construction project requires the use of land outside the site boundaries (or permissions from adjacent landowners) clear and early advice must be provided to the client.[3]

5.10 An early appraisal of the capabilities of the site (that is, what can reasonably be constructed upon the site) must be undertaken with reasonable skill and care even though the design is at a very early stage. The client will often rely upon such early appraisals to take important decisions and to obtain funding.[4]

5.11 The failure adequately to investigate and consider the ground conditions of the site is a frequent cause of claims against engineers and, to a lesser extent, architects. The foundations of any construction must be adequate to bear the loads imposed by the construction and to do so with an appropriate tolerance. The design of foundations able reliably to bear particular loads is something that has to be judged on the basis of what is known about the ground bearing properties of the ground upon which the foundations will be imposed. Whilst conventional foundation design incorporates factors of safety, as a general rule deeper and more reliable foundations cost more and take longer to construct. Assessing the right foundation design is therefore a matter of judgement which involves both a clear appreciation of the ground conditions and an accurate appraisal of the loads which will be imposed.

5.12 Responsibility for an accurate assessment of the ground conditions may thus fall to three different sets of construction professionals: the civil and structural engineer may have to make an assessment of the surveys and investigations which are required; the specialist geotechnical engineer may have to carry out those surveys and provide an analysis and the architect or project manager may have to ensure that the civil and structural engineer and the specialist geotechnical engineer are properly coordinating their efforts and that the design of the civil and structural engineer take proper account of cost considerations.[5]

---

2 In *Acrecrest Ltd v WS Hattrell & Partners* [1983] QB 260 (CA), architects were held negligent in giving insufficient regard to the removal of trees from a site for development. Water which would have been absorbed by the trees was absorbed by the ground, giving rise to swelling or "heave" of the clay, and damage to the new building. In *Kaliszewska v John Clague & Partners* (1984) 5 Con LR 62 an architect designing foundations for a new bungalow who failed to appreciate the significance of removal of trees from a site based on London clay was found negligent.

3 In *John Harris & Partnership v Groveworld Ltd* [1999] PNLR 697 (TCC), architects failed to notice that part of the proposed site was owned not by their client but by the local authority. They were liable for increased costs incurred in obtaining a revised planning permission.

4 In *Gable House Estates Ltd v The Halpern Partnership* [1995] 48 Con LR 1 architects carelessly overestimated the amount of office space which could be obtained from a development. The client purchased property in the belief that it could obtain a particular return based upon the inflated figure.

5 Cases involving a failure adequately to consider ground conditions are numerous and go back at least as far as *Moneypenny v Hartland* (1824) 1 Car & P 351 where the architect mistakenly assumed that the soil beneath the planned foundations for a bridge over the River Severn was solid rock. For a more recent example of case which is concerned with the analysis of the results of ground investigations and the appropriateness of a particular foundation design see *Cooperative Group Ltd v John Allen Associates Ltd* [2010] EWHC 2300 (TCC), where the issue was whether a competent engineer could have advised that vibro compaction ground improvement works could provide an adequate basis for the foundation design of a supermarket.

SECTION B: PLANNING AND REGULATION

5.13 Planning consultants are not treated as construction professionals for the purposes of this work, but architects are frequently concerned with planning applications and project managers and any construction professional undertaking contract administration duties needs to be au fait with the planning permission for the development in respect of which it is engaged and to have a reasonable understanding of what planning permissions require.

5.14 Nor is planning the only set of regulatory restrictions which affect the way in which development is undertaken. Regulation, both in the form of building regulation and environmental regulation is an increasingly important aspect of modern construction projects.[6] As with delegation to specialists, the construction professional is expected both to recognise and appreciate the importance of all of these forms of regulation and to recognise and appreciate the limitations of its own expertise so that, in an appropriate case, the client can be advised to obtain specialist assistance.

5.15 Until relatively recently it was part of the architect's function in respect of all kinds of construction project to advise the client as to planning and to act as the client's agent in respect of seeking planning permission. That remains the case for the vast majority of small and medium sized projects, but recent years have seen a trend for planning applications to be undertaken by specialist consultants working closely with the project architects or for the project architect to be part of a multi-disciplinary practice which includes specialists in planning.

5.16 If the obtaining of planning permission is within the architect's scope of services it will be expected to familiar with the process of planning applications, to understand the operation and effect of the various local and national policies and to engage in such consultation and discussion with the planning authority as would be advised or undertaken by a reasonably competent architect holding itself out as being able to offer planning services.

5.17 The precise standard of care to be applied is a matter of some difficulty (a fact which speaks to the problem created by an emergent profession). In *Middle Level Commissioners v Atkins Ltd*[7] a multidisciplinary practice was said to have failed adequately to consider the need for a Certificate of Proposed Lawful Use. Akenhead J grappled with the difficulty of assessing the standard of care to be applied in these terms:

> As was accepted in argument, there is currently no recognised profession of planners as such, although it is clear that many planners are qualified professionals and very experienced in planning matters. For instance, many architects and surveyors have expertise and experience in planning, as do some engineers. However, there are experienced planners employed by local authorities and other firms, some (and perhaps many) of whom may not hold other professional qualifications but are extremely experienced in planning. This therefore is a case in which one must recognise that there is a range of qualifications and/or experience against which the Court must seek to judge whether the Atkins planners fell below the standard to be expected of reasonably careful planners.

5.18 In the subsequent case of *Elvanite Full Circle Ltd v AMEC Earth & Environmental (UK) Ltd*,[8] Coulson J agreed and added:

> There is currently no recognised profession of planners as such. There is no right or wrong way to make a planning application. Moreover, a planning consultant cannot guarantee success;

---

6 The impact of the Construction Design and Management Regulations 2015 has been considered elsewhere.
7 [2012] EWHC 2884 (TCC) at [33].
8 [2013] EWHC 1191 (TCC) at [177].

generally, he will only be liable for damage caused by advice which no planner who was reasonably well-informed and competent would have given.

5.19 Older cases show a reluctance on the part of the courts to find that a construction professional has been negligent in respect of planning applications.[9] This probably reflects the fact that planning is often as much an art as a science. However, that reluctance cannot be anticipated in cases involving a failure to comply with or take note of existing planning restrictions. Here the construction professional is far more likely to face a successful claim.

5.20 In *BL Holdings Ltd v Robert J Wood & Partners*,[10] an architect was instructed to design an office of under 10,000 sq.ft. so that an office development permit was not required. The building designed included a car park and caretaker's flat which did take the building over 10,000 sq ft, but the planning officer of the council said that these were not designated as office space. The council granted planning permission. In fact the statement was wrong, and the council had no power in law to grant permission. The architect was held liable in negligence for acting on the statement and for failing to advise his client that the planning permission might be wrong in law and of no effect. Gibson J held (harshly) that the architect could not rely upon the advice of the planning officer and should have formed his own contrary view. The architect successfully appealed,[11] but the Court of Appeal did not say that mere reliance upon planning officers was a sufficient discharge of duty.

5.21 Plainly any construction professional engaged in design will be expected to be familiar with relevant building codes and ensure that its design is compliant. Often these obligations dovetail with the designer's CDM obligations, many building regulations being directed at, or relevant to, health and safety. With reference to quality of materials they are frequently prescribed by both domestic[12] and European standards.[13] The most important are the Building Regulations 2010 which are relevant to most types of building work.

5.22 Compliance with Building Regulations in construction is generally the primary responsibility of the contractor, but the inspecting architect or engineer will generally be responsible for taking reasonable care to see that compliant materials are used and (on a more practical level) that the construction is ready to receive approval from the inspector.

SECTION C: BUDGETS AND ESTIMATES

5.23 Construction professionals may be asked to provide estimates at any stage of a contract, but as a generality the most important estimates are those provided at the outset. Estimates as to what a project will cost, how long it will take and what capabilities/attributes it will ultimately possess are likely to be critical to the client's decisions, including the decision whether or not to proceed at all.

5.24 Traditionally estimates of this nature were provided by the architect, possibly assisted by a quantity surveyor.[14] That is still the norm in small to medium-sized projects.

---

9 See for example, *Hancock v Tucker* [1999] Lloyd's Rep PN 814. More recent cases tend to turn on their particular facts – see *Fitzroy Robinson Ltd v Mentmore Towers Ltd* [2009] EWHC 1552 (TCC) where it was held that architects were entitled to rely upon the advice of specialist planning consultants engaged by the client.
10 (1979) 10 BLR 48.
11 (1980) 12 BLR 1.
12 British Standards.
13 For example, the Construction Products Regulation.
14 As was the case in *Nye Saunders and Partners v Alan E Bristow* [1987] 37 Build LR 92.

In more substantial projects cost estimates are usually produced by project managers, who may draw upon the expertise of a number of members of the team initially engaged on behalf of the employer.

5.25 Moreover, it will be noted that estimates produced in such circumstances are generally a great deal more sophisticated than a mere account of what the project will cost. They tend to be reports containing a bank of information which can be used not merely for forecasting overall costs, but for forecasting the way in which costs in respect of specific parts of the project will accrue over time (so as to provide cash-flow information) and how they will fluctuate according to different contingencies.

5.26 This information is consistently updated as the project proceeds. It enables the client (or more particularly the client's team)

- to undertake value engineering (the process of finding the best balance between design utility and cost);
- to undertake tender review (enabling the client both to assess the merits of tenders and to negotiate on their contents); and
- to undertake cost control (the process of keeping costs within budget).

5.27 The authorities dealing with estimates are relatively, perhaps surprisingly, few given the importance of estimates to construction projects. As is discussed below, there are a number of practical reasons which, it is suggested, have the consequence that cases involving allegations of negligent estimation are relatively uncommon.

5.28 It goes without saying that the process of producing a budget or estimate must be undertaken with reasonable skill and care. However, absent some plain and straightforward oversight or miscalculation, it will be a rare case in which a construction professional steps outside the generous band of reasonable decisions in any particular instance.

5.29 Cost estimating is the province of cost advisers, generally quantity surveyors. However, it is important to keep in mind that whilst other construction professionals may not be responsible for providing cost estimates, they may be responsible for providing advice which either takes into account, or requires the obtaining of, costs estimates.

5.30 At the start of any project the client is likely to ask "what will it cost?". It will be a rare case where such advice is neither expressly nor impliedly sought. The function of project managers or lead architects is to enable the client to make the best approximation it can in answering that question. In *Riva Properties Ltd and ors v Foster and Partners Ltd*,[15] the architects had agreed to provide services corresponding to RIBA stages A–L. The court held that stage A required them to identify constraints on the project. One of these was cost. On the facts the client believed it could complete the project for a total budget of under £100 million. This was unrealistic as the architects knew from the cost information which had been provided by costs consultants. They were in breach of duty in failing to give the client that advice.

---

15 [2017] EWHC 2574 (TCC). Fraser J observed:

I consider that whether it is or not should always be identified by an architect exercising reasonable care and skill. Indeed, constraints can only be identified if the existence or otherwise of a budget is established. Mr Rich stated that in relation to most projects (particularly commercial projects) cost is almost always "fundamental" and a "critical and key" constraint, but that is essentially common sense in any event. Constraints must be ascertained and considered by the architect in Stage A, and confirmed in Stage B.

5.31 In this context it should be noted that whilst estimation of cost will generally be central to the client's decision-taking process at an early stage, estimation of time is also likely to be of critical importance. In most construction projects time can be equated with cost, or at least has a close relationship with cost. The longer works take to bring to completion the more expensive they will be. However, time is relevant to other considerations which may bear heavily on the client's mind. These include funding and funding arrangements, cash flow and the client's strategic plans. Just as it will be a rare case in which the client does not expressly or impliedly require advice about cost, so it will be a rare case in which it does not expressly or impliedly require advice about time.

5.32 Estimating is, by its very nature, an uncertain business. It is not akin to calculating the loads on a pillar or the correct paint to use on a particular type of steel. It is process which requires the input of a number of opinions from construction professionals. Whilst the quantity surveyor may be able to make an estimate of quantities from outline design drawings, it will require input from the architect as to material types, from the engineer as to foundation variables and from the project manager as to matters such as contract type. The number of variables leaves plenty of room for error before one even starts to consider that the construction market at the date when the project is put out to tender may be very different to the market in current conditions. Even project estimations undertaken on more modest projects by an architect doing its best on the basis of outline design and an appreciation of likely costs for the particular type of building based on experience are likely to fall within a broad range within which reasonable and competent architects could disagree.

5.33 It is probably no accident therefore that in those rare cases where construction professionals have been found to be negligent the cause of the error has been either elementary miscalculation (something which requires no expert judgment) or a fundamental error of approach.

5.34 In *Nye Saunders and Partners v Alan E Bristow*,[16] the architect was found to be negligent because it failed to allow for inflation at a time of very high inflation. In *Gable House Estates v The Halpern Partnership*,[17] the architect's negligence was not the extent to which its estimate differed from the estimates which might have been produced by the hypothetically competent architect, but its failure to warn the client that it was simply not possible to provide an estimate with the kind of accuracy which the client believed was being provided. In contrast to other functions undertaken by construction professionals, it is particularly difficult to infer from the fact of even a very large differential between outcome and estimate that something must have gone wrong with the process of estimation.[18]

---

16 (1987) 37 Build LR 92 (CA).
17 (1998) 48 Con LR 1.
18 In *Copthorne Hotel (Newcastle) Ltd v Arup Associates (No. 1)* (1996) 58 Con LR 105, the claimant alleged that the defendants must have negligently underestimated the piling costs which turned out at £975,000 against an estimate of £425,000. However, it was unable to produce a figure as to the competent pile diameter. HH Judge Hicks QC said:

> I hope and believe that I am not oversimplifying it if I record the impression that the claimant's main hope was that I would be persuaded to find in their favour by the size of the gap, absolutely and proportionately, between the cost estimate and the successful tender. The gap was indeed enormous. It astonished and appalled the parties at the time, and it astonishes me. I do not see, however, how that alone can carry the claimant home. There is no plea or argument that the maxim "res ipsa loquitur" applies. Culpable under-estimation is of course one obvious explanation of such a discrepancy, but far from the only one.

5.35 The second unusual feature of claims based on alleged negligent under-estimates is that the claimant will often struggle to prove that it has suffered a loss. By the very nature of the task undertaken, the claimant will generally find out about the underestimate once it committed to the project and there is no economic prospect of refusing to continue. It is very difficult for a claimant in those circumstances to show that it has suffered any loss.

5.36 This can be illustrated by an example. Suppose an employer retains a project manager to advise it as to the outturn costs of small factory development. The project manager estimates the outturn costs at £4 million. In fact, the outturn costs are £6 million. The error arose because the project manager omitted to the fit out costs which were part of the project cost. The employer would not have proceeded had been given a competent estimate. However, what is the loss? If the employer had not proceeded it would not have spent the £6 million, but it would not have the benefit of the new factory. Absent some very unusual circumstances, the new factory will be worth the £6 million he spent on it.

5.37 In the rare situation where the client discovers the error before it is too late to pull out (or simply runs out of money because of the additional cost) the employer will be able to claim the wasted costs which it has incurred. It is does not appear that damages have in fact been awarded on this basis in recent times.[19]

5.38 There is authority to suggest that a construction professional who gives an negligent under-estimate is not entitled to recover any fees for his work. In the antique case of *Moneypenny v Hartland*[20] Best CJ held that:

> if a surveyor delivers in an estimate, greatly below the sum at which work can be done, and thereby induces a private person to undertake what he would not otherwise do, then I think he is not entitled to recover.

This appears to have been the basis upon which the architect in *Nye Saunders* was unable to recover its fees. Notwithstanding that the rule has been followed, it is suggested that, simply stated, it is not good law. If the client is entitled to recover its wasted expenditure, it should recover the cost of paying fees as damages in the normal way. If the construction professional's work was worthless a client *may* be relieved of the obligation to pay fees on the basis that there has been non-performance.[21] However, absent either of those situations there is no juridical basis for depriving the construction professional of his fees.

5.39 Plainly these remarks apply to claims by employers against construction professionals. The situation of claims by third parties may be very different. If a construction professional provides an estimate to a bank or other lending institution, either pursuant to a contract or in circumstances where the construction professional owes a tortious duty of care to take reasonable care in making the estimate, the third party may suffer substantial loss by reason of the construction professional's negligence and may be awarded damages in respect of that loss. In particular if the construction professional's negligence leads a

---

19 This may have been the basis of the award in *Dalgleish v Bromley Corporation* [1953] 162 EG 623, but the report is obscure.
20 (1826) 2 Car & P 378.
21 See *William Clark Partnership Ltd v Dock St PCT Ltd* [2015] EWHC 2923 (TCC).

lending institution to advance monies when it would not otherwise have done so (or more monies when it would have ceased making advances) the construction professional may be liable for all of that loss.[22]

SECTION D: DESIGN

Introduction

5.40 Design is one of the core functions of the architect and engineer. Indeed, particularly for architects it is often regarded as the most important function.

5.41 The success with which the obligation to carry out the design of a building or other construction project is generally judged by the efficacy of the finished building or structure, but the process of design may be continual throughout the life of the construction project from the earliest outline design to the last variation. Except in rare cases it will seek to achieve a balance between desirable attributes (utility, strength, durability, aesthetics) and cost. By necessity, the designer will often be dependent upon information provided by others. All of these features of the business of producing a design provide the potential for claims against construction professionals.

5.42 The business of design will generally involve the coordination of the construction professional's design with the design of others.

5.43 Whilst the quality of the construction professional's design may be measured by an assessment of the design as constructed, the way in which the construction professional carries out its design obligations may be very important. The designer will generally be required to produce its design within certain times and in any event in good time to allow the contractor to proceed at an economic pace. Clarity of the design documents is important. The contractor who cannot follow the designer's intentions or, worse still, misinterprets those intentions will proceed more slowly and at greater cost. The designer needs to be able to respond to queries with reasonable speed and, where called upon to modify its design for the client's reasons, to do so clearly, diligently and with due regard to the consequences for the overall efficacy of the design.

Whose obligation?

5.44 Many cases involving construction professionals are not concerned with the fact that the construction professional has undertaken its design its design poorly, but with the fact that it has not undertaken a particular design at all.[23]

5.45 As explained in Chapter 2, modern construction projects frequently involve a complex web of interlocking obligations where, amongst other responsibilities, responsibility for design can be parcelled out, or shared, between the architect or engineer, the specialist sub-consultant and the specialist contractor. In this environment disputes may occur as to who had responsibility for a particular aspect of the design. Ultimately the resolution of

---

22 Subject to considerations of scope of duty – see Chapter 3.
23 In *Board of Trustees of National Museums and Galleries on Merseyside v AEW Architects and Designers Ltd (No1)* [2013] EWHC 2403 (TCC), the architects incorrectly believed that the design of the steps seats and terraces was entirely the responsibility of the specialist reinforced concrete sub-contractor. In fact there were architectural features which also fell within the scope of the services which the architects had agreed to provide.

these disputes is a matter of contract: the express or implied terms of the construction professional's appointment should determine the extent of its responsibility.

5.46 However, the contractual position may not always provide the complete answer. A construction professional which has voluntarily assumed responsibility for the design of part of the building or structure may owe the client (or some third party) a tortious duty of care to see that that design was competent.

5.47 Responsibility to provide design must be contrasted with responsibility to review design. This is dealt with below.

The standard to be achieved

5.48 As discussed in Chapter 4, the almost universal standard to be achieved is the standard of reasonable skill and care. Very few engagements of architects or engineers will require a different standard of performance for design than is required for other aspects of the service provided by the construction professional.[24]

5.49 One important exception is where a client requires the construction professional to produce a design which performs to an objective standard.[25]

5.50 Another is where construction professional acts as a "one stop shop" for a package of services required by the client. Very often multi-disciplinary practices will agree to provide clients with a complete design package, because they have the resources to call upon the skills of (say) both architects and engineers. In other cases, a firm of architects may offer clients a complete design package because they are able to sub-contract the engineering design to engineers or specialist aspects of the design to specialist sub-contractors.

5.51 In such instances, absent some very clear wording to the contrary in the contract, the client will be entitled to expect that the design is carried out to the standard of the reasonably competent designer working in the relevant discipline. Thus it will be no defence for the architect to contend that it acted reasonably in engaging the services of what it thought were competent engineers if the engineering design is one which no reasonably competent engineer would have produced.

Achieving the client's objectives

5.52 The obligation of the designer is to produce the building or structure which the client wants.

5.53 At its simplest level this means following the client's instructions. If the client asks for the bath to be made of pink glass the architect will act in breach of duty if, disregarding

---

[24] There have been suggestions in some of the authorities that a designer might be required to produce a design which is fit for the purpose for which it was intended (regardless of its competence). The decision in *Greaves and Co (Contractors) Ltd v Baynham Meikle and Partners* [1975] 1 WLR 1095 (CA) is sometimes said to support that proposition. It is suggested that this is a mistaken view, both of the law in general and of *Greaves* (which appears to rest upon the view of the court that there was a common understanding which could be elevated to an implied term). The better view is that very clear words are required to require a professional person to perform to a higher standard than that of reasonable care and skill.

[25] This was the court's construction of the warranty given by the floor designers in *Holland Hannen & Cubitts (Northern) Ltd v Welsh Health Technical Services Organisation* (1985) 35 BLR 1, although the basis upon which the majority in the Court of Appeal arrived at that view is open to question.

this instruction on aesthetic grounds, it designs a bath of more conventional material and colour.[26]

5.54 In practice, the obligation to achieve the client's objective means something else: it means designing the building or structure so that it can perform in the way that the client is entitled to expect. The factory where the loading bays are too narrow for the client's vehicles is thus likely to result from a failure to design with reasonable skill and care. Similarly the shopping centre where the atrium restaurants are unusable in winter because the automatic doors at both ends of the building cause cold draughts, is probably poorly designed. The supermarket floor which cracks so that trolleys cannot run across it smoothly may not have been damaged as a consequence of design errors, but design errors in the design of the foundations and the slab are a strong possibility.[27]

5.55 Of course, there will be circumstances in which the client's objectives cannot be fully met and they are not met because of conscious design decisions. The most obvious example of this is where the building the client wants cannot lawfully be constructed because of planning restrictions. It may not be possible to construct a new housing development with eight dwellings because the planning restrictions with regard to highway access mean that the client's chosen means of access cannot be used.

5.56 Another limiting factor is cost. The client may wish to construct eight dwellings but may only have the finance to build six. In these circumstances the designer needs the client's approval to produce a design which does not meet the original objectives. In the absence of that approval the designer may produce a competent design, but may still have acted in breach of its obligations.

5.57 This illustrates an important aspect of the way in which a designer produces its design: knowledge of the client's objectives is not generally sufficient; the designer will often need to consult with the client on the way in which, and the limitations to, the achievement of those objectives.

5.58 As has been set out above, there are rare cases where the effect of a particular engagement is to require the construction professional to go further than this. It follows that in the usual case, the fact that the building as constructed contains flaws does not, of itself, mean that the designer fell below the standard of reasonable skill and care. This is so even where the client establishes that the contractor faithfully followed the designer's design and that no third party has played a part in bringing about the problem. That said, and subject to the qualifications set out below, it will be a rare case in which the designer who cannot advance a clear explanation succeeds.

---

26 In *Gilbride v Sincock* [1955] 166 EG 129, the architect was liable to the employer because the bedrooms were smaller than shown on the plans which the employer had approved. There was no flaw in the design (the bedrooms were perfectly serviceable as bedrooms), but the design did not correspond with the client's intention.

27 Such cases may be thought to contain obvious errors. More difficult situations arise where the perceived flaw in the design is only an error by the designer if it should have known that the outcome was likely to be unacceptable to the client. In *Stormont Main Working Men's Club and Institute Ltd v J Roscoe Milne Partnership* (1988) 13 Con LR 127, the design of the snooker hall did not allow unrestricted access to all of the tables because of the position of some of the pillars. The tables were generally playable (and there was no evidence of substantial difficulty) but the claimant asserted that the designer should have been aware of the requirements of the Billiards and Snooker Control Council. The claim failed because, as a matter of fact, the designer had not been informed that the client wanted the snooker hall to be suitable for competitive snooker (which might be thought a narrow distinction).

5.59 The client's objectives should not be considered merely in terms of the functioning of the completed structure. There will be few clients who wish to pay more for a construction project than is really necessary and an important aspect of the designer's brief is to produce a design with an appropriate cost. The shape and dimensions of the structure, the choice of materials and the processes which are necessary to construct the design all have cost consequences.

5.60 Some design decisions will be axiomatic because of these considerations (for example where there are two building materials of equal efficacy, one of which is substantially cheaper than another). Others will involve questions of judgement. A common example of the latter is the decision as to what kind of foundation design will be employed in respect of building. An engineer who specifies deep piling may achieve a foundation which carries an extreme factor of safety, but the client's interests will not be served if piling was unnecessary because some other cheaper ground improvement scheme would have been effective. It is right to note that it is rare for a client to succeed against a client in an allegation of "over design"[28] (not least because the band of "reasonable" choices may be quite broad), but in situations where a cheaper alternative is simply overlooked there is no reason why a breach of duty will not be found to exist.

5.61 Moreover, the designer should have an eye on practicality. A design which is difficult for the contractor to construct is likely to prove to be unnecessarily expensive. His Honour Judge Newey QC encapsulated this point in *Equitable Debenture Assets Corp Ltd v William Moss Group*[29] when he said:

> I think that if the implementation of part of a design requires work to be carried out on site, the designer should ensure that the work can be performed by those likely to be employed to do it, in the conditions which can be foreseen, by the exercise of the care and skill ordinarily to be expected by them. If the work would demand exceptional skill ... then the design will lack what the experts in evidence described as "buildability". Similarly, I think that if a design requires work to be carried out on-site in such a way that those whose duty it is to supervise it and/or check that it has been done will encounter great difficulty in doing so, then the design will again be defective. It may perhaps be described as lacking "supervisability".[30]

5.62 The level of detail required in any particular design will depend upon the circumstances, but in any event the design must be clear so that the contractor can follow it without difficulty. In *Catlin Estates Ltd v Carter Jonas*[31] the court held that the designer's duty was to "communicate the design to the contractor in a manner which was clear and unambiguous".

Specific design attributes

5.63 A building or other structure may be competently designed in that it performs in the way the client could reasonably expect, but it may still contain flaws which result from poor design.

---

28 For example of failure see *London Underground Ltd v Kenchington Ford and Ors* [1998] 65 Con LR 1.
29 (1984) 2 Con LR 1 at 21.
30 In *George Fischer Holding Ltd v Multi Design Consultants Ltd* (1998) 61 Con LR 85, the designers of a roof were held to have produced an unrealistic design because it specified a series of panels laid end to end and joined by end laps instead of being continuous from eaves to ridge. The consequence was that the design was very difficult to construct and in the event proved not to be watertight.
31 [2005] EWHC 2315 at [340].

5.64 *Safety*. Safety occupies a special place in the consideration of competent design because it is the paramount attribute in respect of which all others are subservient. The designer is obliged to take care to avoid even the smallest risk. In *Eckersley v Binnie*,[32] the issue was whether the defendant engineers should have guarded against the small chance that methane would accumulate in the tunnel in such a way as to present a risk of explosion. Both the court at first instance and on appeal held that they should. Lord Reid said:

> It does not follow that, no matter what the circumstances may be, it is justifiable to neglect a risk of such small magnitude. A reasonable man would only neglect such a risk if he had a valid reason for doing so: e.g. that it would involve considerable expense to eliminate the risk. He would weigh the risk against the difficulty of eliminating it … A person must be regarded as negligent if he does not take steps to eliminate a risk which he knows or ought to know is a real risk and not a mere possibility which would never influence the mind of a reasonable man.

5.65 As is set out above, the statutory overlay of the Construction (Design and Management) Regulations provides an additional reason why safety should be the designer's first priority. In theory the designer who perceives a risk may discharge its duty to its client by providing the client with clear and emphatic warnings. However, whether the giving of such warnings would be sufficient to prevent the designer being liable to some other person who might be injured if the risk manifested itself must be open to doubt. There may be situations where the designer's only option is to refuse to provide a design which carries a risk of injury or death.

5.66 Where safety may be an issue there is an enhanced need to ensure that design is clear and that the contractor appreciates what it needs to do so as to ensure that a safe structure is constructed. In *Bellefield Computer Services & Others v E Turner & Sons Ltd & Others*,[33] Potter LJ stated:

> It is reasonable to expect that in a matter which affects the design safety of a building, whenever the architect has reason to suppose that the achievement of a detail of construction which is not illustrated upon the contractual drawings but which it is necessary should be effected in a particular manner or to a particular standard, may present problems of interpretation and solution to the contractor or sub-contractor who is to effect the work concerned, there is an obligation upon the architect, before the work is done, to clarify with the contractor or sub-contractor his design intention and/or the solution to be employed. This will usually be best achieved by the supply of a detailed drawing. However, depending upon the nature of the detail or technique required, and the skill and understanding of the contractor or sub-contractor, it may be sufficient to resolve the matter by written instructions, by express approval of the contractors' or sub-contractors' proposed solution, or in direct discussions which render the matter clear.

5.67 *Suitability of materials*. Plainly materials which fail shortly after installation cast question on the competence of the design. This may result in the designer becoming liable to the client even where the designer has relied upon the advice of others as to the properties of the material used.[34]

5.68 However, what of materials which perform adequately at the point of completion, but which subsequently deteriorate or fail? A building which suffers substantial rust or which becomes leaky may not be competently designed even though it performs as the

---

[32] (1988) 18 Con LR 1.
[33] [2002] EWCA Civ 1823.
[34] See, for example, *Richard Roberts Holdings Ltd v Douglas Smith Stimpson Partnership (No. 2)* (1988) 46 BLR 50.

client wanted. Durability and sustainability are thus important attributes in any structure and there is a substantial body of published material, both in the shape of codes/standards and other guidance, which is directed to ensuring that designers (generally the architect) select materials which enable the building to have a "design life" of 25, 50 or even 75 years with a reasonable level of maintenance. Other guidance is provided as to materials which present health hazards (for example, because they are flammable).

5.69 As a general rule, failure to follow commonly accepted guidance will be strong evidence of a breach of duty. There may be occasions where a designer wishes to depart from the recommended materials, but to do so requires good reason and (probably) prior consultation with the client.

5.70 *Sustainability*. A building may both look beautiful and perform in all the ways in which the client wanted, but it may be ruinously expensive to heat. That "failing" in the building may be a design flaw which may have resulted from a failure by the designer to employ reasonable skill and care. Whilst the client's immediate objectives may have been to produce a building with certain characteristics within a certain cost, the client would probably have said, if asked, that it also wanted a building that was (reasonably) economic to run. The competent designer would have consulted the client at the outset, giving the client the choice between certain characteristics (often aesthetic) and likely levels of long term cost. In recent times, the energy efficiency of buildings has become better researched and substantial guidance now exists, some of it incorporated into building regulations, but energy efficiency is just one aspect of the overall consideration which is that account should be taken of the long term costs of the building or structure which is being designed.

Coordination and integration of design

5.71 The complexity of modern construction projects means that there is high premium on the coordination of design. Thus the structural design undertaken by the engineer has to marry up with the architectural design prepared by the architect when it comes to the positioning and shape of walls and pillars. This design in turn has to be coordinated with the design of the services engineer who may wish to take ductwork to the roof from a room in the basement.

5.72 The process of design coordination will generally be under the overall supervision of the architect who is usually asked to act as "lead designer" or "lead consultant" or "design team leader" and the process of design consultation will be monitored by the project manager, who will ensure that backlogs do not build up and that failing consultants are harried into performance.[35]

5.73 The mechanics of design coordination are usually the circulation of design through various iterations (the architect's drawings being provided to the civil and structural engineer, its drawings being produced having had regard to those of the architect, the mechanical and electrical engineer's drawings being produced having had regard to both of those sets of design and its drawings being circulated to the other designers). Design changes by any one designer (unless very minor) will be the subject of consultation with the other designers, often at regular design team meetings.

---

35 For examples of cases where construction professionals were held liable for failing adequately to coordinate their designs with the designs of others see *Equitable Debenture Assets Corp Ltd v William Moss Group Ltd* (1984) 2 Con LR 1 and *Royal Brompton NHS Trust v Hammond (No. 9)* [2002] EWHC 2037 (TCC).

5.74 The obligation to coordinate design does not bring with it an obligation to assess the competence of the design produced by others. Absent some very obvious flaw, the architect is not expected to assess the viability of the design of the civil and structural engineer and vice versa.[36] However, it is suggested that each designer must understand the design of others sufficiently to be able to assess, at least in broad terms, how it will impact on its own work.

5.75 At its simplest level this means the avoidance of clashes, so that drawings of the different designers are at least compatible as to what goes where. At a more evolved level the designers (and particularly the lead designers) will need to have at least an understanding of the performance specification to which other designers are working. If the civil and structural engineer knows that the architect is designing a supermarket floor which is suitable for the use of supermarket trolleys it should appreciate that the extent of acceptable differential settlement may be limited by flatness requirements. If the architect knows that the mechanical and electrical engineer is proposing a comfort cooling system with a particular number of air changes over a particular period it should appreciate that there are likely to be significant restrictions upon any variation to the architectural design which provides an increase in natural ventilation.

5.76 In *Board of Trustees of National Museums and Galleries on Merseyside v AEW Architects and Designers Ltd (No. 1)*,[37] the architects were held to be negligent when they failed to appreciate that the lighting design, which involved a raised lighting track, made it less likely that the ceiling tiles (which were being installed by a specialist sub-contractor) would be adequately fitted.

Novel design

5.77 It is sometimes said that special considerations apply to the production of novel design and in particular design which uses untested materials or techniques. In the Victorian case of *Turner v Garland and Christopher*[38] Erle J directed the jury as follows:

> You should bear in mind that if the building is of an ordinary description, in which [the architect] had had an abundance of experience, and it proved a failure, this is an evidence or want of skill and attention. But if out of the ordinary cause, and you employ him about a novel thing, about which he has had little experience, if it has not had the test of experience, failure may be consistent with skill. The history of all great improvements shows failure of those who embark in them; this may account for the defect in the roof.

5.78 However, the client whose project is ruined because the design fails is unlikely to be comforted by the knowledge that it has played its part in the history of all great improvements, and there is no special rule which applies in situations involving novel design: the *Bolam* test is sufficiently flexible to take account of the situation of the designer who chooses to employ techniques or materials which, by definition, are not employed by any substantial body of the profession.

5.79 Where a designer undertakes an unusual design or makes use of some untried procedure there will always be an obligation upon the designer to justify that course. Generally

---

36 See *Holland Hannen & Cubitts (Northern) Ltd v Welsh Health Technical Services Organisation* (1985) 35 BLR 1 (CA).
37 [2013] EWHC 2403 (TCC).
38 (1853) Hudson, Building Contracts (Fourth Edition), Vol. 2, p. 1.

the justification is the novel or challenging nature of the project or some aesthetic imperative. It is implicit in the need for justification that the design is rational and that the designer has taken reasonable care to anticipate potential risks.[39] The real question is how the *Bolam* test applies where the designer has justification for the novel design, but something goes wrong which causes the design to fail.

5.80 It is suggested that such cases raise three particular questions: (1) has the loss or failure occurred because of the novelty of the design? (very often the problem will be a humdrum error which happens to occur in the course of the novel design); (2) if it has, is the manifestation of a risk which was capable of being anticipated (if only in general terms)?; and (3) if so, was the client given an adequate warning about that risk? What is an adequate warning will depend upon all the facts.

The obligation of the reviewer

5.81 One of the most difficult areas of design responsibility for the construction professional is the function of reviewing the design of others.

5.82 It is very common in substantial construction projects for elements of the works to be designed by sub-consultants or specialist sub-contractors. The architect or engineer will usually be accorded a reviewing obligation in respect of that design. This will require the construction professional to exercise the reasonable care and skill of an architect or engineer in considering the design of the specialist. In most cases the architect or engineer will not have the same degree of specialist expertise, but it will have sufficient expertise to spot obvious mistakes, to see whether the design has been coordinated with other design and to check that the designer has followed recommended practice.

5.83 Thus, the civil and structural engineer will often have a review obligation in respect of the design of the specialist groundworks contractor or concrete slab erector. The architect will often have a review obligation in respect of the curtain wall specialist subcontractor. The mechanical and electrical engineer may have a review obligation in respect of the design of the air conditioning specialist subcontractor.

5.84 The obligation will often have two correlating facets. Because specialist sub-consultants and sub-contractors generally produce their design pursuant to specifications which are produced by the architect or engineer, the level of protection afforded to the client will often be affected by the quality of the specification. The more detailed and the clearer the specification, the less likely it is that the specialist will fall into error.

5.85 The second facet concerns the interrogation of design. There may be an onus on the architect or engineer to seek clarification from the sub-consultant or sub-contractor of important aspects of the design even where there is no indication of any mistake or failure to follow good practice. This is particularly the case where the architect or engineer lacks the specific expertise to ascertain how the design will cope with known risks. Thus the engineer considering the specialist foundations design may wish to investigate the extent to which the design allows for variability of ground conditions across the site, the architect reviewing the curtain walling design may wish to query how a particular gasket is intended to prevent unwanted water ingress into the structure, the mechanical and electrical engineer

---

39 See *Eckersley v Binnie* (1988) 18 Con LR 1 (CA).

examining the design of the air conditioning may want to inquire as to how the design copes with a mix of large and small rooms.[40]

Inspection or supervision by the designer

5.86 Historically building contracts were administered by the architect who was also the designer. Early standard forms of architects' contracts tended to fuse the inspection function, undertaken by the architect performing its contract administration role, with a *supervision* function, performed by the architect carrying out its design obligations.[41]

5.87 The inspection role accorded to the contract administrator is now performed by any construction professional competent to take on that work (see below) and rarely involves any element of supervision. Contracts between clients and construction professionals with design obligations in the context of design and build contracts will often include not merely an obligation to review the contractor's design, but often an obligation to review the contractor's works (to ensure that the design is being constructed properly).

5.88 The precise ambit of this obligation will depend, in the first instance, on the wording of the obligation as it appears in the appointment, but the meaning of the words may be coloured by the nature of the works and what might reasonably be required.

5.89 Straightforward construction works, which do not require any special skill taking them out of the ordinary course, are unlikely to require close attention even when a novel design is being constructed. By contrast, highly skilled work which is essential for the success of the design (for example, the installation of a particular form of plasterwork in a historic building) may require the designer to "stand over" the contractor, if only on a sample instance, to ensure that method of working is satisfactory.

5.90 The same may be true for critical aspects of the design where the contractor has to use a particular product or carry out its works in a particular way.[42] Moreover, special attention may be merited when substantial design responsibility has been accorded to the sub-contractor. Thus it may be necessary for the civil and structural engineer to observe the way in which the specialist piling sub-contractor carries out the piling in order fully to discharge its obligation to review the specialist sub-contract works.

Approval of defective design

5.91 There is an important distinction between consultation with a client in which the construction professional outlines the risks inherent in a particular design, and consultation where the construction professional merely puts the design in front of the client to obtain

---

40 For an example of such considerations see *Try Build Ltd v Invicta Leisure (Tennis) Ltd* (1997) 71 Con LR 140 where civil and structural engineers were held to be liable to the employer for failing adequately to review the design of the specialist sub-contractor. An element of the failure consisted in the fact that that the engineer's specification was weak. Another was that the engineers did not sufficiently challenge the specialist sub-contractor as to how aspects of its design were intended to perform.

41 See the 1962 RIBA Conditions of Engagement discussed in *Sutcliffe v Chippendale & Edmondson* (1982) 18 BLR 149.

42 On one view this was the reason for the liability of the engineers in *Ministry of Defence v Scott Wilson Kirkpatrick and Dean & Dyball Construction Ltd* (reported only on appeal at [2000] BLR 20). The contractors had not followed the method of fixing required by the contract and the roof was blown off in a storm. The importance of the fixing was such that the engineers should have checked it was being employed.

its approval. The latter will rarely afford the construction professional with any defence. For client approval to exculpate the construction professional the client must be fully and adequately informed. The less sophisticated the client, the clearer the explanation or warning will need to be.

5.92 The fact that a design may have been given building regulations approval as effected does not mean that it was competent.[43] Even if the building regulations inspector was competent and diligent, the competence of the design is to be judged by the court according to the standard of care to be expected of a competent designer. It is unlikely that the opinion of any person other than an expert would affect the court's judgment. In practice, even as a forensic argument, the contention that the design must have been competent because the building regulations inspector or some other qualified person did not complain about it could only apply when the alleged defect would have been obvious.

5.93 Moreover, the fact that some other professional person in the client's team approves or agrees with a defective design does not make that design any less defective. In *Pride Valley Foods Ltd v Hall & Partners (Contract Management) Ltd (No. 1)*[44] His Honour Judge Bowsher QC said:

> A competent architect does not present a design that he knows to be deficient in an important respect and then discuss with the client whether the deficiency should be removed. Still less does he present such a design and say, I did not need to tell the client about the deficiency because the client already knew that such a feature was required. Take a simple example. An architect designs a house as a residence for a client who happens to be a surveyor and forgets to require a damp proof course under a parapet wall. If after construction the client complains, it is no answer for the architect to say, "Well you knew about the need for a damp proof course as well as I did". The architect is employed to use his own skill and judgment. There is no duty on the part of the client who happens to have a particular skill to examine the architect's designs and tell the architect where he has gone wrong. If I as a lawyer go to a solicitor for advice and pay him for it, I do not see why I should be criticised if I fail to do that solicitor's work all over again and check whether he has got it right.

5.94 Nor will it be a defence to an allegation of breach of duty that the problem arose because of a late design change demanded by the client. Late design changes can present significant risks to the coordination and integration of design[45] and (at the very least) the designer faced with such an instruction should provide a strong warning to the client unless it is absolutely satisfied that no risk is presented in that case.

SECTION E: THE TENDER PROCESS

Introduction

5.95 Almost without exception, construction projects pass through a tender process during which (1) the employer's advisers prepare the package of documents setting out the proposed design and proposed contractual terms; (2) different contractors consider that package of documents in order to decide whether and if so at what price they will offer

---

43 See *Sahib Foods Ltd (in liquidation) v Paskin Kyriades Sands* [2003] EWHC 142 (TCC).
44 (2001) 76 Con LR 1.
45 See for example *Leach v Crossley* (1985) 30 BLR 95 (CA), where the engineers failed to follow though the consequences of a late change to the design of a car park.

to carry out the works and (3) the preferred or selected contractor negotiates final terms (which can and often do include documents outlining the way in which the contractor proposes to complete the project – the "contractor's proposals").

5.96 All of the employer's team of construction professionals will be involved in this process.

5.97 The design team will be involved in preparing and (in so far as necessary) revising the design and specifications for the works. They will address queries raised by contractors and evaluate the merits of design proposals and methodologies advanced by the selected or preferred contractor.

5.98 The architect, engineer or project manager taking lead responsibility for the project will advise the client as to the course of the tender process, including (if necessary) as to the choice of contract and when and how this is to be agreed.

5.99 The quantity surveyor will be responsible to advising as to matters of cost. In respect of contracts which make use of a bill of quantities it will be the quantity surveyor's function to draw this up. In respect of other kinds of contract it will be the responsibility of the quantity surveyor to make estimates of what particular aspects of the work should cost so that the client can evaluate the tenders provided by different contractors. The quantity surveyor and other members of the client's team may be involved in assessing and costing revisions suggested by the contractor.

5.100 Because all construction contracts are concerned with the balance of risk they will all have a part to play in advising the client as to the merits of counter-proposals by contractors (for example as to design responsibility).

Specification

5.101 As set out above, part of the process of design is the provision of buildable and economic plans and specifications to the contractor. Particularly on larger more complex projects the role of the designer will include addressing technical queries raised by contractors and evaluating counter proposals. Whether design risk is to be retained by the client or transferred to the contractor the efficiency and accuracy of the clarifications and advice provided by the design team in this process will translate into cost and time.

Bills of quantities

5.102 Bills of quantities are less common than they were when traditional contracts were the norm, but they still play an important role, particularly for contracts which are likely to be costed according to volumes of material or rates for certain types of work. The quantity surveyor will be expected to be abreast of market trends for pricing purposes and to have a thorough understanding of construction practices. The quantity surveyor who makes an error in the drawing up of a bill of quantities may cause its client loss. Whilst obvious and plan mistakes may not bind the client (for example, leaving out a decimal point), more subtle mistakes are likely to leave the client with an onerous obligation.

5.103 In modern times the responsibility of the quantity surveyor at this stage is to assist the architect and (if different) project manager in advising the client as to cost. A client which expects to be able to construct a project for a budget that includes a £10 million contractor's price is not well served if it goes through an expensive and time consuming tender process only to find that no reputable contractor would consider undertaking the project for less than £15 million.

5.104 The quantity surveyor will need to have a keen appreciation of way in which the design will be constructed in practice. If the constraints of the site mean that particular work has to be carried out in a particular way that fact will be reflected in the quantities. The quantity surveyor will generally report to the client through the medium of the architect, engineer or project manager (who can be expected to identify points of concern), but again clarity and efficiency are important attributes. The quantity surveyor which fails to report clearly on a certification issue may cause the client loss just as if it had reported incorrectly.

Selection of contractors

5.105 The client will look to the professional team and specifically the project manager, architect or engineer with lead responsibility for advice on the selection of contractors.

5.106 This is not just a matter of choosing the contractor who provides the lowest price. The client's interests are not well served by engaging a contractor which puts in an unreasonably low tender or which misunderstands some aspect of the project. That contractor may become tempted to cut corners on an onerous contract or indeed may become insolvent. The more experienced and sophisticated the client, the less guidance the project manager will need to provide.

5.107 In so far as advice on selection of a contractor involves matters of judgment, the band of reasonable advices which a contract administrator might provide is likely to be a broad one. However, the project manager which offers no or no substantial advice when advice is required or which overlooks an obvious indication that a contractor is or might be unsuitable may be vulnerable to a claim.

5.108 If and in so far as the employer is engaged in the selection of nominated sub-contractors, the selection of a sub-contractor which lacks the skills or resources to carry out the work effectively risks making the employer vulnerable to a claim by the contractor. A particular danger, which may not be apparent to employers, is that it may be difficult clearly to apportion responsibility for a particular event between the contractor and the nominated sub-contractor. As with advice on the selection of contractors, the band of reasonable advices may be broad, but a failure to advise as to obvious risks could be a breach of duty.

5.109 On large projects the solvency of contractors is particularly important and it may be part of the scope of the project manager's duties to advise the client as to the need to obtain parent company guarantees and/or performance bonds. It is no part of the project manager's duty to investigate the solvency of potential contractors, but particularly where the employer is or appears to be inexperienced it probably is part of the project manager's duty to raise the issue of protection against insolvency.

Construction contracts

5.110 The project manager and/or lead construction professional may be required to advise the client as to the form of contract. This is a difficult area for construction professionals as the meaning and effects of contracts are matters of law.

5.111 On small to medium sized contracts it is generally the responsibility of the architect (as the lead construction professional) to advise the client as to the type of contract which best meets the client's needs and as to how and in what ways such contracts work. As is set out above, there are a number of different suites of standard form contracts with which architects, engineers and quantity surveyors will expected to be familiar. Unless the client indicates that it does not require advice, the construction professional will be expected to suggest a form of contract, explaining why that form has been selected and why it offers advantages over other forms of contract.

5.112 It is very unlikely that a construction professional could ever be liable to a client for advising selection of a contract from one suite rather than another, but the construction professional which advises that a management contract be employed when a building contract with a contractor's design portion would have been more appropriate could be in breach of duty.

5.113 On larger projects clients will be advised by specialist solicitors who are likely to spend long periods negotiating those contracts with lawyers appointed by the contractors. Even here, however, the project manager or lead construction professional is likely to have important responsibilities. The client's lawyers may be best placed to explain to the client how any particular form of contract works (and to draft bespoke clauses dealing with particular risks), but the project manager or lead construction professional will have the knowledge of the project and the industry experience to explain the operation of any particular provision in practical terms and to enable the client to evaluate and price particular risks.

5.114 Because it is common for larger projects to commence on site before contracts are agreed, construction professionals are often involved in advising clients as to the use of (and even drafting) letters of intent. Plainly it is important that the construction professional understands not just the limitations of this form of agreement, but also the risks which a series of letters of intent may present to the client.[46]

Other contracts

5.115 It should be noted that, particularly on larger projects, the construction contract is not the only contract which the client needs to effect.

5.116 Part of the role of the project manager may well be to ensure that the client has obtained the correct insurance (that is, is a named beneficiary under a policy obtained by the contractor), has the benefit of collateral warranties from the design team if their contracts are being novated and (at latest by the end of the project) is provided with all guarantees and sub-contractor warranties which it is the contractor's obligation to provide.

5.117 Some of these obligations may properly fall within the sphere of "contract administration" (see below), but some of them necessarily go beyond the mere competent administration of the building contract.

---

46 In *Trustees of Ampleforth Abbey Trust v Turner & Townsend Project Management Ltd* [2012] EWHC 2137, project managers were held to be liable to clients for failing to advise as to the risks presented by the utilisation of series of letters of intent which led to the work being completed without a formal contract and without the contractual protections which would have been afforded to the client had the project manager insisted that the contractor turn its attention to agreeing the contractual terms.

5.118 In *Pozzolanic Lytag Ltd v Bryan Hobson Associates*,[47] engineers were appointed as project managers for a project involving the design and construction of a storage facility for pulverised ash. The contract required the contractor to take out insurance against the risk of a particular type of damage to the works. This insurance was not obtained and when the damage occurred the employer was unable to recover its losses from the contractor. The engineers were held to be liable to the client. Dyson J said:

> If a project manager does not have the expertise to advise his client as to the adequacy of the insurance arrangements proposed by the contractor, he has a choice. He may obtain expert advice from an insurance broker or a lawyer. Questions may arise as to who has to pay for this. Alternatively he may inform the client that expert advice is required, and seek to persuade the client to obtain it. What he cannot do is simply act as a "postbox" and send the evidence of the proposed arrangements to the client without comment.[48]

SECTION F: CONTRACT ADMINISTRATION

Introduction

5.119 Architects, engineers, quantity surveyors and project managers commonly perform contract administration roles in construction projects. The precise obligations of the roles will depend upon the terms of the contract which is being administered (and these are always the first place to ascertain what it is that the construction professional is required to do). In many forms of contract the contract administrator is given a different title (for example "project manager", "architect" or "engineer") and the scope of services will vary very considerably. However, a typical appointment will require the construction professional to perform certain core tasks :

- liaising with the contractor (acting as the client's agent for all communication with the contractor)
- periodically inspecting the works
- giving instructions, including variation or change orders
- determining any applications for extensions of time by the contractor
- authorising interim payments to the contractor
- certifying the date of completion
- settling the adjusted contract sum (final account).

5.120 "It is the Contract Administrator's duty to administer the project in such a way that the correct contractual procedures and good administrative practices are followed, and that the life of the building contract from inception to completion is accurately and completely recorded. Where specific procedures do not exist the Contract Administrator should carefully consider what activities ought to be undertaken and how they are communicated in order to ensure that the project is fairly and effectively administered. The Contract Administrator should devise and promulgate procedures that will address this. The

---

47 [1999] BLR 267.
48 At [272].

conditions of contract are intended to regulate not only written matters that arise during the course of the building contract, but also set out the parties' agreement as to how they will resolve potential disputes. The building contract conditions should not be regarded as an intention to seek conflict and the Contract Administrator should recognise the part that can be played in achieving the completion of a project in a manner satisfactory to both the employer and the contractor. The proper administration of the building contract by the Contract Administrator is one way in which this can be achieved".[49]

5.121 Whilst the role of contract administrator only exists during the period of the contract (including any defects liability period) it is the norm for the contract administrator to be a construction professional which is already engaged by the client in some other capacity. Whereas there are standard form appointments which are limited to the role of contract administrator, these tend to be used infrequently. Project managers are generally appointed at an earlier stage in the project and contract administration will form a part of their agreed duties.

5.122 It is common for a construction professional to share the responsibility of contract administrator and designer (architect and engineer), or contract administrator and cost consultant (quantity surveyor). Thus the construction professional's responsibilities to the client will probably commence in a period before the construction contract is entered into and may be more extensive than is required for mere good administration of the contract.

5.123 Not least because of the importance of the certification role of the contract administrator, it remains common for the contract administrator be a named person. As is set out above,[50] this does not mean that a construction professional will assume a personal liability for the carrying out of the contract administrator's functions, but it does restrict the construction professional's ability to delegate tasks and the ability of any firm or company to operate the role of contract administrator through a team.

Familiarisation

5.124 It goes without saying that a competent contract administrator needs to familiarise itself with the project (if not already familiar), its participants and the contract. Familiarisation with the contract is particularly important. Construction contracts tend to be highly detailed and pedantic in their requirements. There are generally specific procedures which must be followed in respect of design approvals, design changes, payment applications, extensions of time and so forth. On larger projects the client and its lawyers will often have negotiated a large number of specific additions and deletions to the standard form in use. Specifications and addenda may contain important provisions. The contract administrator which is not fully on top of the contract provisions may fail adequately to carry out its obligations and may cause the client loss.[51]

---

49 RICS Guidance Note, Contract Administration.
50 See Chapter 3.
51 See, for example, *West Faulkner Associates v Newham LBC* (1994) 71 BLR 1, where the architects misunderstood the provisions for serving a notice stating that the contractor was failing to proceed regularly and diligently. Brown LJ said that the architect "was required to have a general knowledge of the law as applied to the most important clauses ... of standard forms of building contract".

Providing information and giving instructions

5.125 In the perfect construction project the contract administrator never has to give information or instructions to the contractor: the design is perfectly complete requiring no embellishment or elaboration; it remains entirely unchanged, not least because the client never changes its mind on anything and the project is entirely problem free so that the contractor never needs to consult with the contract administrator over anything. In the real world none of these things happen and a critical part of the contract administrator's role is the provision of information and instruction to the contractor so that the imperfect project can proceed to completion as smoothly (and cheaply) as possible.

5.126 Indeed many construction contracts are predicated upon the continual flow of information between the contract administrator and the contractor (in particular, in connection with the approval of design). For most building contracts design is fixed at the point of contract in the sense that it exists to the level of detail necessary to enable the client to know what it is going to obtain and the contractor to know what it has to price, but it is not fixed in the sense that it is not fully detailed.

5.127 In contracts where the client retains design responsibility, detailed design is released in stages appropriate to the progress of the works. In contracts where design responsibility is provided to the contractor, the latter will generally have to release its detailed design in a similar fashion so that it can be approved by the employer. Even if not a member of the design team, the contract administrator has a key role in facilitating this process.

5.128 The contract administrator should use reasonable skill and care to ensure that appropriate procedures are in place for the transmission and evaluation of design and should closely monitor the operation of those procedures. Many contracts will provide for detailed programmes which identify which design is required and when and will set out specific requirements for queries to be addressed. However, in the absence of such provisions, it is almost certainly an implied term of any construction contract that the employer will provide any design it is obliged to provide within a reasonable time having regard to the contractor's requirements[52] and it follows that it is the obligation of the contract administrator, as the client's agent, to use reasonable skill and care to ensure this happens.[53]

5.129 Nor are these the only circumstances in which instructions will be required as a matter of routine. Others include:

- additions, omissions or variation to the works
- clarification of the contract documents if sought by the contractor and (in appropriate circumstances) where suggested by other members of the client's team or thought necessary by the contract administrator;
- instructions to place orders with specialist subcontractors;
- instructions for the expenditure of provisional sums;
- instructions for specific tests; and
- opening of completed works so as to verify compliance with the building contract.

---

[52] See *Neodox Ltd v Borough of Swindon and Pendlebury* (1977) 5 BLR 34.
[53] The position in respect of the contractor may be different. Many contracts require contractors to work to programmes or to respond within certain times to reasonable requests for information. However, in the absence of these, the contractor generally only has an obligation to complete the works by a certain date.

5.130 The contract administrator will be required to deal with matters which fall outside the anticipated process of the supply and/or approval of design information. Unexpected problems may emerge on site. These may require an instruction from the client. Depending on the precise balance of risk which the parties have accepted, the client may be required to provide information and to do so within a reasonable time.[54] These instructions may amount to variations. Alternatively, decisions taken by the client for other reasons may result in design changes which are variations.

5.131 The exercise of reasonable care and skill by the contract administrator may require more than the mechanical processing of such matters. In a proper case it may be necessary for the contract administrator to advise the client as to further inquiries, costs consequences and potential future problems. In may also require the contract administrator to liaise with different members or the design team and/or the contractor.

5.132 Aside from acting with care in appreciating the need for such instructions and acting with diligence to provide them, it is important that the contract administrator takes care in the way in which they are conveyed. Most construction contracts will provide that instructions for variations be given in writing (there is often a complex code for costing them) but in emergency situations important instructions, including for variations, may be given orally. Care in both the clarity of the instruction and in keeping a clear record of the instruction will be expected of the competent contract administrator. It is generally good practice to confirm oral instructions as soon as practicable after they have been made, if not in a formal instruction document then in a site meeting which is minuted.

Certification

5.133 "Certification" is the process whereby the contractor obtains a formal decision or approval which determines the contractor's entitlement to payment or relief from some penalty and which the contract requires be provided in a fair manner. The bargain between an employer and a contractor as contained in modern construction contracts is replete with certification. There are:

- interim or progress certificates, which determine the contractor's entitlement to specific parts of the overall contract price when particular stages in the progress of the works have been reached;
- final certificates which indicate that work has been completed (sometimes deeming that it has been completed to a particular standard) and which set out the balance of the sum due to the contractor less any specified retention;
- certificates of completion, either marking the fact that all of the works are substantially complete or that specific parts of the works are substantially complete, thus providing a benchmark against which the contractor's liability in respect of completion is ascertained;
- certificates extending the time for completion where the contractor has or is likely to be delayed for a reason which is not its risk;
- certificates of non-completion marking the point at which the contractor is in breach of its obligation to complete by the completion date (as extended); and

---

54 See *Merton LBC v Stanley Hugh Leach Ltd* (1986) 32 BLR 51.

- miscellaneous certificates triggering potential entitlements on the part of the employer, for example certificates of failure to remove defective work.

5.134 The process of certification is generally undertaken by the contract administrator. Indeed certification is one of its most important functions. It will be immediately apparent that the accordance of this role to the employer's agent (rather than some neutral third party) gives rise to an apparent conflict between the bargain between the employer and the contractor (that the certification process will be fair) and the bargain between the employer and the construction professional (that the construction professional will act in the best interests of the employer). How is that conflict to be resolved and in what circumstances will a construction professional, acting as certifier, be liable to the employer for loss caused by the way in which it has carried out certification?

5.135 The courts have resolved this difficulty by reading the obligation of the contract administrator to its client as being shaped by and (ultimately) subservient to the obligation of the employer to provide the contractor with a fair certification process. Construction contracts generally create a distinction between the agency role of the contract administrator (where it acts as the client's representative) and the decision taking role (where it acts, not as arbitrator, but none the less from a position of detachment).[55]

5.136 In *Sutcliffe v Thackrah*[56] the House of Lords considered that this special role was not one that aligned an architect with arbitrators or experts who were immune from suit, but it did shape the extent of the architect's duty to his client so that it could only be sued in narrow circumstances. Lord Reid said:[57]

> It has often been said, I think rightly, that the architect has two different types of function to perform. In many matters he is bound to act on his client's instructions whether he agrees with them or not, but in many other matters requiring professional skill he must form and act on his own opinion. Many matters may arise in the course of the execution of a building contract where a decision has to be made which will affect the amount of money which the contactor gets. Under the RIBA contract many such decisions have to be made by the architect and the parties agree to accept his decisions. For example, he decides whether the contractor should be reimbursed for loss under clause 11 (variation), clause 24 (disturbance), or clause 34 (antiquities), whether he should be allowed extra time (clause 23) or when work ought reasonably to have been completed (clause 22). And, perhaps most important, he has to decide whether work is defective. These decisions will be reflected in the amounts contained in certificates issued by the architect. The building owner and the contractor make their contract on the understanding that in all such matters the architect will act in a fair and unbiased manner, and it must therefore be implicit in the owner's contract with the architect that he shall not only exercise due care and skill but also reach such decisions fairly, holding the balance between his client and the contractor.

5.137 On the basis of *Sutcliffe* and other authority, Jackson J in *Scheldebouw BV v St James Homes (Grosvenor Dock) Ltd*[58] summarised the role of the certifier in as follows:

> 34 ... In many forms of building contract a professional person retained by the employer, and sometimes a professional person directly employed by the employer,

---

55 Indeed this distinction is not confined to construction contracts. It can arise in any contract where the parties have elected to clothe the agent of one of them with a decision taking role. In the shipping case *Panamena Europa Navigacion Compania Limitada v Frederick Leyland & Co Ltd* [1947] AC 428, the House of Lords distinguished between the agency role of the owner's surveyor and his decision taking role.
56 [1974] AC 727 (HL).
57 At 737.
58 [2006] EWHC 89 (TCC).

has decision-making functions allocated to him. I will call that person "the decision-maker". The decisions which he makes are often required to be in the form of certificates, but this is not always so. For example, there are many contracts (of which the present one is an instance) in which extensions of time do not take the form of certificates.

35   Three propositions emerge from the authorities concerning the position of the decision-maker.

   (1)   The precise role and duties of the decision-maker will be determined by the terms of the contract under which he is required to act.
   (2)   Generally the decision-maker is not, and cannot be regarded as, independent of the employer.
   (3)   When performing his decision-making function, the decision-maker is required to act in a manner which has variously been described as independent, impartial, fair and honest. These concepts are overlapping but not synonymous. They connote that the decision-maker must use his professional skills and his best endeavours to reach the right decision, as opposed to a decision which favours the interests of the employer.

36   In my judgment, these propositions are all applicable to the construction manager in the present case. The fact that the construction manager acts in conjunction with other professionals when performing his decision-making function does not water down his legal duty. When performing that function, it is the construction manager's duty to act in a manner which is independent, impartial, fair and honest. In other words, he must use his professional skills and his best endeavours to reach the right decision, as opposed to a decision which favours the interests of the employer.

5.138 It can be seen from this analysis that the following broad propositions govern certification and the obligations of the construction professional as certifier.

5.139 Where a construction professional accepts an obligation to perform some decision taking role in a construction contract (that is, making determinations as between his client and the contractor) the construction professional's implied obligation to serve the interests of his client must be read subject to its role in the construction contract.

5.140 Under the construction contract the construction professional carrying out decision taking will be obliged to act in a matter which is independent, impartial, fair and honest.

5.141 This does not necessarily involve the provision of a certificate (it is the substance of what the construction professional which is required to do which matters, not the form).

5.142 Thus it will not be open to the client to instruct the construction professional to act in a way which is not independent, impartial, fair and honest (and if a client persists in such an instruction the construction professional may have to consider resignation).

5.143 But this does not mean that the construction professional is released from all obligation to its client.

5.144 In the first place, whilst acting in a manner which is independent, impartial, fair and honest it must also act carefully. Honesty, independence and impartiality will be no defence to an allegation of carelessness.

5.145 In the second place, whilst the client is not permitted to require the construction professional to provide a certificate which the construction professional does not believe is independent, impartial, fair and honest, that does not mean that the construction professional is entitled to override an instruction not to provide any certificate at all. That kind of instruction may make the contractual machinery inoperative, but it does not require the construction professional to contravene honesty, independence and impartiality.

5.146 There are a plethora of authority where the courts provide guidance on the exercise of particular certification functions in particular construction contracts.[59] These cases are really concerned with the contractual provisions rather than the competent execution of the certification process and each case turns upon the precise wording of the clause in the building contract under which the obligation to certify arose. They are concerned with the obligations of the certifier to the contractor in his capacity as the employer's agent, not the obligations of the certifier to its client.

5.147 It should be noted that whilst there are few decisions concerning alleged negligent certification[60] this may be more to do with the employer's contractual rights rather than any difficulty in suing contract administrators. In most construction contracts certificates or equivalent decisions are rarely final in the sense that they can generally be opened up in a subsequent adjudication, arbitration or litigation.[61] Moreover challenges to the accuracy of certificates generally involve a decision on the merits of the certificate rather than the reasonableness of the contract administrator.

5.148 It follows that, in the usual case, an employer dissatisfied with a certificate may take legal proceedings against the contractor to recover any loss caused by a certificate being erroneous. It is only in those cases where (unusually) there is some legal bar to recovery or where the contractor is insolvent that there would be a need to take action against a construction professional.

Inspection

5.149 A construction professional will often be required to inspect works. The obligation generally arises in one of two ways.

5.150 In the most common instance the construction professional will be required to inspect work as part of its contract administration role. The point of the inspection will be to assess the stage at which the work has reached (for example, whether it is substantially complete) or to inspect work which the contract administrator is concerned is or might be defective. Plainly there is some overlap between an assessment of the completeness of work and its adequacy and inspections addressing adequacy may involve other construction professionals if the contract administrator lacks the expertise to make an assessment.

---

59 The best-known concern decisions granting or refusing extensions of time: *Perini v Commonwealth of Australia* [1969] 12 BLR 82; *Merton LBC v Stanley Hugh Leach Ltd* (1986) 32 BLR 51; *Balfour Beatty Building Ltd v Chestermount Properties Ltd* (1993) 62 BLR 12; *Henry Boot Construction (UK) Ltd v Malmaison Hotel (Manchester) Ltd* (1999) 70 Con LR 32; *Royal Brompton Hospital NHS Trust v Hammond & ors (No. 7)* [2001] EWCA Civ 206; *City Inn Ltd v Shepherd Construction Ltd* [2010] BLR 473; *Walter Lilly & Co Ltd v Mackay* [2012] EWHC 1773 (TCC).

60 For an example of a claim which failed, see *Royal Brompton Hospital NHS Trust v Hammond (No. 1)* [1999] BLR 162.

61 Indeed interim certificates can often be reviewed during the life of the contract and corrected in the final account.

5.151 The second way in which the obligation arises is where the construction professional is the designer of works and undertakes an obligation to the employer to review the construction of those works to ensure adherence to the design (see above). In rare case, the designer may be required to supervise works, but this obligation is very unlikely to be imposed upon the contract administrator.

5.152 The contract administrator's appointment may contain provisions which stipulate the number of times that the contract administrator may be required to visit site (implicitly, to inspect works). Such obligations are generally framed in terms of a minimum frequency of attendances because the precise number and duration of attendances will depend upon what is reasonably required in order to allow the contract administrator adequately to perform its function.[62] If works are progressing smoothly on a straightforward project it may only be necessary for the contract administrator to inspect works at the stages appropriate for interim payments and other contractual milestones.

5.153 By contrast if a project encounters difficulties, particularly difficulties which threaten to cause substantial delay or cost, it may be necessary for a contract administrator to attend almost daily during periods of crisis and to inspect critical aspects or the work.[63]

5.154 Guidance from the courts tends to focus on cases where the contract administrator (usually an architect) has failed to observe defective or non-compliant work. Each case tends to be determined on its special facts, a helpful analysis concerning some of the relevant considerations was provided in *McGlinn v Waltham Contractors Ltd*[64] where HH Judge Coulson QC assessed the legal principles relating to an architect's obligation to inspect:[65]

a) The frequency and duration of inspections should be tailored to the nature of the works going on at site from time to time ... Thus it seems to me that it is not enough for the inspecting professional religiously to carry out an inspection of the work either before or after the fortnightly or monthly site meetings, and not otherwise. The dates of such site meetings may well have been arranged some time in advance, without any reference to the particular elements of work being progressed on site at the time. Moreover, if inspections are confined to the fortnightly or monthly site meetings, the contractor will know that, at all other times, his work will effectively remain safe from inspection.

b) Depending on the importance of the particular element or stage of the works, the inspecting professional can instruct the contractor not to cover up the relevant elements of the work until they have been inspected ... However, it seems to me that such a situation would be unlikely to arise in most cases because, if the inspecting officer is carrying out inspections which are tailored to the nature of the works proceeding on site at any particular time, he will have timed his inspections in such a manner as to avoid affecting the progress of those works.

---

62 See *Corfield v Grant* [1992] 29 Con LR 58 where HH Judge Bowsher QC explained that the frequency of inspection depends upon what is happening on site from time to time.
63 For example, the laying of a damp proof membrane or any other important feature which may become inaccessible when further works are undertaken.
64 [2007] EWHC 149 (TCC).
65 At [281].

c) The mere fact that defective work is carried out and covered up between inspections will not, therefore, automatically amount to a defence to an alleged failure on the part of the architect to carry out proper inspections; that will depend on a variety of matters, including the inspecting officer's reasonable contemplation of what was being carried out on site at the time, the importance of the element of work in question, and the confidence that the architect may have in the contractor's overall competence…

d) If the element of the work is important because it is going to be repeated throughout one significant part of the building, such as the construction of a proprietary product or the achievement of a particular standard of finish to one element of the work common to every room, then the inspecting professional should ensure that he has seen that element of the work in the early course of construction/assembly so as to form a view as to the contractor's ability to carry out that particular task…

e) However, even then, reasonable examination of the works does not require the inspector to go into every matter in detail; indeed, it is almost inevitable that some defects will escape his notice…

f) It can sometimes be the case that an employer with a claim for bad workmanship against a contractor makes the same claim automatically against the inspecting officer, on the assumption that, if there is a defect, then the inspector must have been negligent or in breach of contract for missing the defect during construction. That seems to me to be a misconceived approach. The architect does not guarantee that his inspection will reveal or prevent all defective work … It is not appropriate to judge an architect's performance by the result achieved …

5.155 The following general conclusions can be drawn:

- The starting point for considering the extent of the construction professional's obligation is its appointment.
- The general requirement is to carry out such inspections as are reasonably necessary to confirm that that the work is contractually compliant.[66]
- The mere fact that a contract administrator has failed to observe non-compliant or defective work does not, of itself, imply negligence.[67]
- Even the fact that defective work was observed by the contract administrator does not automatically imply negligence. Thus if defects are concealed it may not be negligent to have missed them.[68]

5.156 However, the need to inspect is enhanced in certain situations:

- where the work is of particular importance (either for contractual milestones or because of its design significance).[69]

---

66 *Sutcliffe v Chippendale and Edmondson* [1982] 18 BLR 149.
67 *East Ham Corp v Bernard Sunley & Sons Ltd* [1966] AC 406 (HL).
68 *Trustees of London Hospital v TP Bennett & Son* (1987) 13 Con LR 22.
69 *Ministry of Defence v Scott Wilson Kirkpatrick* (reported only on appeal at [2000] BLR 20); *Six Continents Retail Ltd v Carford Catering Ltd* [2003] EWCA Civ 1790.

COMMON ISSUES IN A CONSTRUCTION PROJECT

- where the work is unlikely to be able to be inspected subsequently.[70]
- where the contract administrator suspects the work is defective/non-compliant or knows that it is and has given instructions for its remediation.[71]

5.157 On some construction projects the employer engages a clerk of works, part of whose function is to monitor the adequacy of construction works being undertaken on site. The contract administrator may properly rely upon the clerk of works for some functions and to some extent, but it cannot delegate its inspection duties to the clerk of works. This balance is probably best viewed through the prism of what is reasonably necessary. The presence of a clerk of works is one of the facts to be taken into account in deciding what is reasonably necessary but it is not a determinative fact.[72]

5.158 There is no prescribed way of carrying out an inspection and each project will have different needs. However, at a high level of generality, a court or arbitrator may take note of the guidance provided by the RICS which states that:

The inspection should include:

(a) The contract administrator recording basic information, which can usefully be done on a standard sheet noting such matters as:

    date and time of inspection
    weather conditions
    number and type of operatives on site
    basic details of any instructions given on site
    progress of the works, including any specific activities being undertaken during the visit or not yet undertaken

(b) A review of the quality of workmanship related to the contract documents. The contract administrator will need to be familiar with the preliminaries or preambles in the specification and relevant codes of practice, etc.

(c) A review of progress in relation to the contract programme. The contract administrator may need to discuss with the contractor action to be taken to ensure the programme is followed.

(d) A check on materials being used. Instructions may be required for substitute materials.

(e) A check that the works conform to the specification and drawings. The contract administrator may need the assistance of specialist designers of the works (e.g. structural or services engineers) and their inspections should ideally be coordinated with those of the contract administrator.

(f) Noting general information to enable the contract administrator to report to the employer on the progress and quality of the works.

(g) Records of any measurement of work that might be needed for certification purposes, particularly where they may be covered up before the valuation date.

(h) A general awareness of health and safety arrangements on site, with any concerns being brought to the attention of the contractor. It is advisable to confirm the outcome in writing. It can be useful to issue "site directions" at the end of an inspection to confirm any instructions given. It is recommended that the contract administrator issues copies to the employer and contractor's head office upon return to the office. Self-carbonating pads can be useful to facilitate this. The site direction will have the dual purpose of providing a clear record of the instructions given, and satisfying the contract

---

70 *George Fischer Holding Ltd v Multi Design Consultants Ltd* (1998) 61 Con LR 85.
71 *Kensington, Chelsea and Westminster AHA v Wettern Composites* [1985] 1 All ER 346; *McGlinn v Waltham Contractors Ltd* [2007] EWHC 149.
72 *Ministry of Defence v Scott Wilson Kirkpatrick* (reported only on appeal at [2000] BLR 20).

administrator's obligations of confirming instructions in writing. The site direction can then be converted to a formal contract instruction shortly afterwards.

Reporting to the client

5.159 An important part of the contract administrator's function is reporting to the client. Aside from the general obligation to keep the client informed as to progress (for example by provision of minutes of meetings and reports) the contract administrator will have specific obligations to consult with employers in relation to facts coming to the attention of the contract administrator which are likely to have a material bearing on the client's interests. These include:

5.160 *Costs*. The contract administrator may have a specific cost reporting responsibility under its appointment (if only a responsibility to pass on and explain the reports of some other construction professional with a detailed cost reporting brief) but in the absence of any specific obligation it would generally be part of the administration function to appraise the client as to (i) the achievement of cost stages in the works (ii) the extent to which additional costs have been or are forecast to be incurred over and above the original contract sum and (iii) matters which may provide future costs risks or indeed opportunities to save costs.

5.161 *Time*. The contract administrator will be obliged to keep the employer briefed as to progress and the extent to which progress approximates to the contractual intent. In complex projects where the contractor is required to submit programmes for approval, the client will need to be consulted and advised. In the generality of projects it will be the responsibility of the contract administrator to make its own ascertainment of progress (which may not accord with that of the contractor) and to advise the client. If variations are proposed which have substantial time consequences the client will usually be consulted.

5.162 *Disputes and differences*. There are very few construction projects which do not involve disagreements between the employer and the contractor as to time or money or both. There may be other disagreements (for example as to the scope of design responsibility or the adequacy of work). Whilst it is the role of the contract administrator to act for the employer in respect of these matters (sometimes carrying out its certification responsibilities) it should not be forgotten that in all instances the contract administrator acts as the employer's agent.

5.163 Generally the contract administrator will be obliged to keep the client informed as to these disputes or disagreements and often it will be required to provide the client with advice. In giving advice concerning disputes the construction professional must take particular care when advising as to what may be matters of law. Thus whilst a contract administrator will be expected to appreciate the meaning and effects of "pay less notices" (and to issue such notices correctly) it may find that it lacks the expertise adequately to advise the employer in respect of an adjudication. The competent advice in that situation may be that the client should consult solicitors.

Keeping records

5.164 Whilst the appointment may be silent as to the contract administrator's duties in respect of records it seems likely that a court would imply a term into any contract administrator's contract that it should keep reasonable records of the project. The point of keeping adequate records is threefold: they enable the contract administrator better to carry out its function in administering the contract; they enable the client to understand and demonstrate what events took place and when they took place (something which may be important in the context of any dispute with a third party); they enable the contract administrator to demonstrate that it carried out its duties properly. Whilst the contract administrator's obligation to its client may only encompass the first two reasons, the third reason is important. If some problem emerges which causes the client loss and damage the client may well look to the contract administrator for an explanation and (implicitly if not explicitly) reasons why the contract administrator was not at fault.

5.165 Whilst the extent of good practice is variable, depending upon the particular project, a contract administrator would generally be expected to maintain:

- A complete copy of the contract.
- Notes of site inspections: recording the date, time and duration of the visit; the weather, persons on site, progress and status of the works; any defects in workmanship and the action taken; and instructions given, particularly if verbal.
- Copies of all other instructions.
- Minutes of meetings.
- Office files: collated so that information (letters, emails, notices, telephone notes, instructions, drawings etc.) can be easily found and tracked.
- Photographs: filed in a manner so that the location and date taken can easily be identified.

5.166 At a minimum such documents should be kept for six years from the end of the defects liability period. If either the contractor's obligations or the construction professional's obligations are recorded in a deed, it may be advisable to maintain records for 12 years.

CHAPTER 6

Causation loss and damage in claims against construction professionals

Introduction	113
Section A: the essential approach	114
Section B: scope of duty	114
Section C: remoteness	116
Section D: causation	118
Effective cause	118
Break in the chain of causation	119
Outcome dependent upon hypothetical actions	120
Section E: assessment of loss	122
Costs of rectification or diminution in value?	122
Other expenditure or costs	124
Liability to third parties	125
Personal injury, inconvenience and distress	126
The construction professional's fees	127
Section F: mitigation	128
Section G: limitation of actions	129
Section H: contributory negligence	132
Section I: contribution	134

INTRODUCTION

6.1 A construction professional may have acted in breach of contractual and tortious duty without causing the person to whom that duty was owed any loss. Indeed, although a breach of contractual duty will always amount to "liability" even if it results in no damage,[1] there is no breach of tortious duty unless damage is caused, "damage" being an essential ingredient of the tort of negligence.

6.2 A claimant who brings a claim against a construction professional for breach of duty must be able to prove not just that the construction professional acted in breach of duty, but that the breach of duty caused the claimant loss.

6.3 English law has developed a number of approaches and tests which guide the courts in deciding (1) whether loss has been caused and (2) what level of damages to award in respect of that loss.

6.4 Moreover where the claimant has been *partly* responsible for the loss or the breach of duty took place a long time previously the law affords the construction professional a

---

1 Usually attracting a nominal award of £5 damages.

number of partial or even complete defences to having to pay damages. It also provides the construction professional with the opportunity to make others also responsible for the loss contribute to the claimant's damages.

6.5 These are not facets of the law which are peculiar to construction professionals, but they are very important in the arena of professional liability claims against construction professionals.

SECTION A: THE ESSENTIAL APPROACH

6.6 At its most basic, "causation, loss and damage" in claims against construction professionals can be summarised as follows: (1) the claimant having proved (on the balance of probabilities) that the construction professional breached some contractual or other duty owed to it; (2) the claimant must prove (again on the balance of probabilities) that the breach of duty caused the claimant harm; (3) the claimant must provide evidence as to the financial value of that harm and (4) the court will award damages having regard to that harm which, in so far as money can do it, will put the claimant in the position it would have been in had the breach of duty not occurred.

SECTION B: SCOPE OF DUTY

6.7 As the Supreme Court recently observed,[2] the English courts impose restrictions upon the extent to which a claimant who has suffered harm because of a breach of duty can recover the monetary value of that harm.[3] The best known of these restrictions is the rule that no recovery can be made for harm which is too remote (see below). However, in recent times, the courts have emphasised a further restriction which is particularly apposite to breaches of duty by professional advisers: harm which has been caused the giving of negligent advice will only sound in damages if, on a true consideration of the professional's duty, this was the kind of harm which was within its "scope of duty".[4]

6.8 This means that the negligent adviser should not be made liable for all of the consequences of the advisee entering into a transaction, but only the consequences which are appropriate to the responsibility it undertook. In a case involving valuers[5] Lord Hoffman explained the nature of the scope of duty restriction as follows:

> It is that a person under a duty to take reasonable care to provide information on which someone else will decide upon a course of action is, if negligent, not generally regarded as responsible for all the consequences of that course of action. He is responsible only for the consequences of the information being wrong. A duty of care which imposes upon the informant responsibility for losses which would have occurred even if the information which he gave had been correct is not in my view fair and reasonable as between the parties. It is therefore

---

2 *Hughes-Holland v BPE Solicitors* [2017] UKSC 21.

3 ... the law is concerned with assigning responsibility for the consequences of the breach, and a defendant is not necessarily responsible in law for everything that follows from his act, even if it is wrongful. A variety of legal concepts serves to limit the matters for which a wrongdoer is legally responsible...

(per Lord Sumption JSC at [20])

4 A phrase subsequently considered by Lord Hoffman to be better expressed as "scope of liability", but which none the less has gained common currency despite its imprecision.

5 *South Australia Asset Management Corp v York Montague Ltd* [1997] AC 191 (HL) at 214. The "scope of duty" restriction is invariably referred to as "SAAMCO".

inappropriate either as an implied term of a contract or as a tortious duty arising from the relationship between them.

The principle thus stated distinguishes between a duty to provide information for the purpose of enabling someone else to decide upon a course of action and a duty to advise someone as to what course of action he should take. If the duty is to advise whether or not a course of action should be taken, the adviser must take reasonable care to consider all the potential consequences of that course of action. If he is negligent, he will therefore be responsible for all the foreseeable loss which is a consequence of that course of action having been taken. If his duty is only to supply information, he must take reasonable care to ensure that the information is correct and, if he is negligent, will be responsible for all the foreseeable consequences of the information being wrong.

6.9 The scope of duty restriction is not part of the law of causation. It operates at a prior stage. It can be understood as an overarching test applied to whether particular damages are recoverable in any particular case, which takes into account as part of its analytical process the questions: "what was it that the construction professional was required to do?", "what was the nature of the harm sustained by the client, including the way in which it arose?" and "what is the relationship between the construction professionals duty and the harm?", before asking "was this the kind of harm that the construction professional undertook to prevent?". Looked at in this way, it is so closely bound up with the rules concerning causation that it often treated as part of the causation analysis,[6] although properly speaking there is no hard distinction between breach of duty and causation.[7]

6.10 The evolution of the scope of duty restriction is in its infancy and its application to cases concerning construction professionals has been infrequent and (it is suggested) not always satisfactory.[8]

6.11 However, it should be noted that whilst demonstrating that the harm fell within a professional's "scope of duty" is a necessary ingredient of almost every claim against a construction professional, it will be a rare case where it is properly arguable that the restriction applies so as to reduce the loss claimed. This is because the paradigm operation of the restriction is in cases where the claimant acts upon the advice of the defendant so as to enter into some transaction or to take some other step. In that paradigm case the defendant's advice is merely on ingredient in the decision which causes the loss and, consequently, it is unjust that the defendant should bear all of the loss. The principle does not apply where negligent advice is the entire cause of the harm. This is the case in most other instances

---

[6] For a more in-depth analysis the reader is referred to *Jackson & Powell on Professional Liability* (Eighth Edition) at paragraph 9–202.

[7] See *Calvert v William Hill Credit Ltd* [2008] EWCA Civ 1427.

[8] In *HOK Sport Ltd v Aintree Racecourse Ltd* [2002] EWHC 3094 (TCC), the court decided that negligent advice from the architects as to the number of seats that would be provided in a racecourse stand did not entitle the claimant to recover all its losses of the transaction. That part of the decision, it is suggested, was correct. The decision that the claimant could not recover for the loss suffered by its loss of opportunity to redesign the racecourse is questionable. Some element of this loss probably was within the scope of duty. Similarly the court's decision in *Earl's Terrace Properties Ltd v Nilsson Design Ltd* [2004] EWHC 136 that gains made by rising prices should not be credited against the developer's loss is inconsistent with the correct approach (which starts from the actual loss and then applies the scope of duty restriction, rather than employing the scope of duty restriction to ascertain the actual loss). In *Try Build v Invicta Leisure* (2000) 71 Con LR 141 engineers argued that they should not be liable for all the financial consequences of negligent certification. That argument failed on the facts. It was part of the engineers' duty to advise as to the quality of the works being inspected. Consequently they were liable for all the financial consequences of the works being poorly constructed. The decision is correct but there is no very full analysis of scope of duty. For an impeccable application of the restriction see *Bank of Ireland v Faithful & Gould* [2014] PNLR 28.

where construction professionals are sued. Consequently, scope of duty is only likely to play an important role in cases where a construction professional gives advice which forms part of the claimant's decision-taking process when entering into a transaction.

6.12 The distinction can be illustrated by two contrasting examples.

6.13 The developer who decides to purchase and develop office premises on the advice of the architect as to net lettable space may be met with the scope of duty restriction if it seeks to claim all of the losses from entering into that transaction. It may very well be that it would not have entered into the transaction if the architect had given the correct advice. In that sense, the architect has caused all of the developer's loss. However, the decision to purchase will have been made on the basis of many considerations, not the least of which will have been an assessment of whether the market was likely to be favourable in the future. Some of these assessments will have been wrong. It is unjust that the architect should bear the entire loss of the transaction.

6.14 By contrast, the housebuilder which is unable to sell its houses at the height of the market because of negligent setting out drawings by the architect is unlikely to face the same difficulties when it advances a claim based upon the profits it would have made. In that situation the architect did not provide the client with either information or advice which formed the basis of the developer taking some action. The architect's breach of duty operated directly to cause the delay, which in turn caused the loss. The market loss *may* be irrecoverable because it is too remote (see below), but it is not irrecoverable because it falls outside the architect's scope of duty. It is true that considerations of whether the market would go up or down were never part of what the architect was engaged to do, but the developer is not seeking to make the architect liable for the consequences of a judgment about the future state of the market: it is seeking to make the architect liable for the consequences of the setting out being wrong.

6.15 That said, the scope of duty principle is in development. There are suggestions in some of the authorities that it is of broader application than only to situations where the client takes a decision based upon advice or information.[9] It remains to be seen whether it can be applied to cases where the client suffers loss because of some other failing.

SECTION C: REMOTENESS

6.16 Loss may be caused by a construction professional's breach of duty but which may be so far removed from the reasonable expectation of the parties that it would be unjust for the court to make the construction professional liable for that loss.

6.17 Thus, for example, a failure of the mechanical and electrical engineer may cause the air conditioning to be defective. This may cause the client's employees to be unproductive. This may in turn cause the client to lose a substantial contract with a customer. The client will thereby lose profit. Although the client's loss of profit may have been caused by the construction professional's breach of duty, it would be unjust to make the construction professional compensate the client for this loss because it is too "remote".

6.18 Damages for breach of contract:

> should be such as may fairly and reasonably be considered either (1) arising naturally, i.e. according to the usual course of things from such breach of contract itself, or (2) such as may

---

9 See the discussion in *Riva Properties Ltd and ors v Foster and Partners Ltd* [2017] EWHC 2574. Although Fraser J found that the principle had no application in that case, he hinted that it might have done so had he not found that the relevant losses were not caused by the breaches of duty.

reasonably be supposed to have been in the contemplation of both parties at the time they made the contract, as the probable result of a breach of it.[10]

6.19 In tort the test is arguably more generous: the claimant may recover provided that the *type* of damage suffered was reasonably foreseeable at the date when the breach of duty occurred.

6.20 On the current state of the law, the tortious test is likely to be of very little relevance in cases concerning construction professionals. The English courts have long regarded the different remoteness tests for contract and tort as anomalous in the context of persons who owe identical duties of care in both contract and tort. In the relatively recent solicitors' negligence case of *Wellesley Partners LLP v Withers LLP*,[11] the Court of Appeal held that in such cases the contractual test should apply.

6.21 Remoteness in both contract and tort is highly fact specific and there is little to be gained by a review of the different factual analyses. The focus is upon the closeness of the relationship between the type of loss suffered and the expectations and knowledge of the parties at the date of the contract or breach.

6.22 Thus (absent some contractual exclusion) there is no rule that prevents a claimant from recovering loss of profit from a construction professional which has caused that loss of profit by its breach of duty. Rather, the true position is that the loss of profit may be too remote depending upon precisely how this loss arises and precisely what could be said to be in the common contemplation of the parties at the date of the contract. An architect whose negligence delays the completion of a factory may, for example, he liable for loss or profits arising from loss of production over the period of delay. Loss of profits arising from the factory owner's inability to win future work as a result of uncertainty as to when the factory might be completed could be too remote.

6.23 In *John Grimes Partnership Ltd v Gubbins*,[12] the employer had engaged engineers to design a road and drainage for a residential development and to obtain approval from the local authority for the road's design. The engineers failed to complete the work by the date agreed. The employer eventually engaged another engineer, who completed the designs and obtained approval 15 months after the agreed date. The engineers successfully claimed against the employer for unpaid fees, but the judge also allowed the employer's counter-claim, which included damages for the decline in value of the development during the period for which its completion had been delayed as a result of the engineers' breach of contract in failing to complete its work on time. The engineers appealed contending that these losses were too remote.

6.24 The Court of Appeal rejected that contention. The engineers had known that the employer was intending to sell the development. They had known that the property market might move up or down. There was nothing in the contractual relationship which limited the extent to which the engineers might be liable for a breach of duty.[13]

---

10 *Hadley v Baxendale* (1854) 9 Ex 341, 354.
11 [2015] EWCA Civ 1146.
12 [2013] EWCA Civ 37.
13 That is, there was no common understanding (albeit not expressed in the words of the contract) that engineers in these circumstances would not be liable for this kind of loss. In *Siemens Building Technologies FE Ltd v Supershield Ltd* [2010] EWCA Civ 7, the Court of Appeal held that the commonly understood background facts might narrow or expand the types of loss which the parties could be said to have within their reasonable contemplation or which would be reasonably foreseeable.

SECTION D: CAUSATION

Effective cause

6.25 The courts have been anxious not to prescribe rules for establishing causation. "Causation in fact" is no more than the assessment of a sufficient connection between two events (typically breach and loss) so that it can be said that one "caused" the other.[14] It is a common sense test. That said, the courts have occasionally applied an informed oversight so as to distinguish between events which *cause* loss and events which are the mere *occasion* for loss.[15] It is not always easy to appreciate the reasons for these distinctions and it is suggested that, particularly now that the scope of duty restriction is established, causation will be large a matter of the application of inference and common sense.

6.26 In *Riva Properties Ltd and ors v Foster and Partners Ltd*,[16] the architects failed to give the claimant proper advice on what the works were likely to cost. The project (a substantial hotel) could not be built because the claimant lacked the resources to build it. The claimant contended that had it been properly advised, it would have caused the project to be subjected to value engineering so that the cost of construction would have been brought down to a manageable level and the development would have proceeded. The claimant contended it had lost the profit it would have therefore made. That claim failed. The Judge found that even if the right advice had been given and value engineering had occurred, the state of the financial markets was such at the time when the claimant would have required funding that the hotel would not have been built in any event.

6.27 However, difficulties can arise where there are a number of possible causes of damage. Here the courts adopt slightly different tests for contract and tort, although the result is much the same.

6.28 In contract, the court looks to see if a particular cause was an "effective" cause of the loss. If it was, it does not matter if there were other effective causes.[17] An effective cause is one where it can be said that "but for" that cause the loss would not have occurred. In *Axa Insurance UK Plc v Cunningham Lindsey United Kingdom*,[18] Akenhead J observed:

> Particularly in construction cases, there can be more than one effective cause of a loss. There is good authority for the proposition that, where there are two effective causes of loss and one of the effective causes is a breach of contract by a defendant, the claimant can recover that loss from the defendant. It will be no defence for the defendant to say that, even if it had not been in breach, loss would have been caused by the other effective cause in any event.

---

14 See Sir Thomas Bingham MR in *Banque Bruxelles v Eagle Star* [1995] QB 375 (CA) at 406. In *Department of National Heritage v Steensen Varming Mulcahy* (1998) 60 Con LR 33, His Honour Judge Bowsher QC noted that construction should be judged from an informed standpoint: The test is what an informed person in the building industry (not the man in the street) would take to be the cause without too microscopic an analysis but on a broad view.

15 See (in the context of professional negligence) *Galoo v Bright Grahame Murray* [1994] 1 WLR 1360 (CA) and in the context of other negligence *Quinn v Burch Bros (Builders) Ltd* [1966] 2 QB 370 (CA).

16 [2017] EWHC 2574 (TCC).

17 *Heskell v Continental Express Ltd and anor* [1950] 1 All ER 1033 (KB), where Devlin J said:

> It may be that the term "a cause" is, whether in tort or in contract, not rightly used as a term of legal significance unless it denotes a cause of equal efficacy with one or more other causes. Whatever the true rule of causation may be I am satisfied that if a breach of contract is one of two causes, both co-operating and both of equal efficacy, as I find in this case, it is sufficient to carry judgment for damages…

18 [2007] EWHC 3023 (TCC) at [265].

6.29 In tort the test is whether one of a number of causes materially contributed to the loss about which complaint is made.[19] In practice this brings about the same result as the analysis provided by Akenhead J in *Axa*, and it is suggested that there is unlikely to be any circumstance where, for causation purposes, it makes a material difference whether the claim is brought in contract or in tort.

6.30 This can be illustrated by example. Suppose the architect and the engineer are both responsible for a negligent design clash. The architect's design would have been satisfactory had it not clashed with the design of the engineer and vice versa. It is no defence to a claim by the employer for the architect to contend that its breach of duty was not an effective cause of the loss because there was another cause, being the negligence of the engineer. The architect's breach of duty was one of two effective causes of the loss.

6.31 Sometimes the court is faced with a number of potential causes of harm without any clear evidence as to which of them is the effective cause. In these circumstances the court must do its best without resorting to merely ranking causes in order of likelihood.[20]

Break in the chain of causation

6.32 The act of a third party will rarely break the chain of causation: "The question which ought to be asked is whether that intervening cause was of so powerful a nature that the conduct of the plaintiffs was not a cause at all but was merely a part of the surrounding circumstances".[21] The presence of contribution rights between defendants make it unlikely that a court would find that an otherwise causative breach of duty lost its causal potency.[22] An example of the strictness of the court's approach occurred in *Siemens Building Technologies FE Ltd v Supershield Ltd*[23]: The defendants were contractors who were responsible for fitting a defective valve which caused a water tank to overflow. This would have caused no problem had it not been for the fact that the drains had been blocked by a third party. The Court of Appeal rejected the contractor's contention that it had not caused the loss.

6.33 There is slightly more scope for a break in the chain of causation caused by the claimant itself. In *Borealis AB v Geogas Trading SA*,[24] the Court of Appeal set out general guidance, which can be summarised as follows:

- the evidential burden lies with the defendant, whilst the legal burden remains with the claimant;
- in order to constitute a break in the chain of causation the true cause of the loss must be the conduct of the claimant rather than the breach of duty on the part of the defendant;

---

19 *McGhee v National Coal Board* [1972] 1 WLR 1 (HL), where Lord Reid said:

> It has always been the law that a pursuer succeeds if he can shew that fault of the defender caused or materially contributed to his injury. There may have been two separate causes but it is enough if one of the causes arose from fault of the defender. The pursuer does not have to prove that this cause would of itself have been enough to cause him injury.

20 *Fosse Motor Engineers Ltd v Conde Nast* [2008] EWHC 2037 (TCC).
21 *Roberts v Bettany* [2001] EWCA Civ 109 per Buxton LJ at [21].
22 See *Linden Homes South East Ltd v LBH Wembley Ltd* [2002] EWHC 536 (TCC).
23 [2010] EWCA Civ 7. See also *Hi-Lite Electrical Ltd v Wolseley UK Ltd* [2011] EWHC 2153 and *Flanagan v Greenbanks Ltd* [2013] EWCA Civ 1702, where the courts came to similar conclusions.
24 [2010] EWHC 2789 (Comm).

- absent very exceptional circumstances, the claimant's conduct will usually need to be unreasonable, although the mere fact of unreasonable conduct will not of itself suffice;
- the claimant's knowledge of the situation he faces as a result of the defendant's breach of duty is very important; the less it knows of the breach and the danger it faces the harder it will be to show that conduct short of recklessness will suffice to break the chain of causation;
- each case is fact sensitive, involving a practical inquiry into the circumstances of the defendant's breach of contract and the claimant's conduct: this is the overarching principle.

6.34 Those principles were broadly applied by Mr Justice Akenhead in *Carillion JM Group Ltd v Phi Group Ltd*[25] where the consulting engineers had contended that the claimant's actions in instituting unsuitable remedial works broke the chain of causation between the claimant's loss and their negligent supervision of the original design. Here the usual availability of contributory negligence inclines courts to be reluctant to find that a break in the chain of causation has occurred.[26]

Outcome dependent upon hypothetical actions

6.35 All analysis of causation is an exercise in conjecture: the court has to ask itself what would have happened if the breach of duty had not occurred. In many, if not most, cases the answer will be straightforward. If the engineer had produced competent calculations the structural pillars would have been adequate to bear their loadings without cracking. If the architect had carried out a competent inspection the defects would have been uncovered. However, in other cases where the counterfactual involves hypothetical actions which are not plain and obvious there may be some difficulty as to how the court should assess causation.

6.36 *Hypothetical actions of the claimant.* The claimant must prove on the balance of probability what it would have done.[27] Thus if the claimant contends that, had it been properly advised, it would have agreed to pay for the costs of an additional survey must satisfy the court with evidence that this is in fact what it would have done. Whilst the claimant's retrospective evidence is not unimportant, the court generally makes an assessment based upon its appreciation of the contemporaneous facts and relevant matters such as

---

25 [2011] EWHC 1379 at [141].

26 Similar arguments also failed in *Riva Properties Ltd and ors v Foster and Partners Ltd* [2017] EWHC 2574 (TCC), although here there was no contributory negligence.

27 This causation analysis may have a bearing on the kind of loss recoverable (see below). In *Cooperative Group v John Allen Associates Ltd* [2010] EWHC 2300 Ramsey J stated that in a claim against a construction professional for negligent design the claimant must establish what would have happened if the construction professional had exercised reasonable skill and care. If the claimant establishes that it would have proceeded with the construction on the basis of a competent design the measure of loss will be the costs of rectifying the design defects giving credit for any higher costs of construction. If the claimant establishes that it would have abandoned the project (for example, because it would have been faced with much higher construction costs) the correct measure of loss is wasted expenditure.

contemporaneous evidence of the claimant's attitude to risk.[28] The claimant which contends that, properly advised, it would have opted for a much more expensive solution may struggle to persuade a court of its case if the evidence shows that at the time it was doing its level best to cut costs.

6.37 *Hypothetical actions of the defendant.* In some cases the court has to inquire into what the defendant would have done if it had acted within the terms of its contract and/or was presented with different information or instructions. Plainly if (as is usually the case) there was only one way in which the defendant could have acted so as to comply with its contractual obligations there will be no issue. In a small number of cases the defendant has a range of options available to it. These cases raise particular difficulties, but the general principles are clear. Whilst commercial parties may be able to contend that they would have acted in the way most favourable to their interests,[29] where a construction professional is called upon to exercise reasonable skill and care the outcome will be determined by an assessment of what the professional would have been most likely to do, having had regard to all the circumstances including its actual conduct.[30] Even though these cases concern the hypothetical actions of the defendant, the burden of proof remains with the claimant.

6.38 *Hypothetical actions of third parties.* It is not uncommon for claims against construction professionals to involve consideration of the hypothetical actions of third parties.

6.39 Usually these situations present no difficulty because the hypothetical action is plain and obvious. If the project manager had been told about the architect's concerns, it would not have permitted works to continue. If the architect had correctly advised as to the quality of the work, the contract administrator would not have certified practical completion.

6.40 However, there are occasions when the answer is not so clear. If the project manager had correctly advised the employer as to the availability of the performance bond the employer may have seek to have drawn upon it, but what would the bondsman have done? If a planning application was made late, what would planning authority have done if it was made with reasonable diligence?

6.41 Where there is any doubt as to what the third party would have done the correct analysis is that set out in *Allied Maples Group v Simmons & Simmons*[31] apply: (i) if the

---

28 In *North Star Shipping Ltd v Sphere Drake Insurance plc* [2005] EWHC 665 (Comm) at [254], Colman J dealt with hypothetical assertions as to what would have happened if further information had been made available to relevant advisors or decision makers:

> it is important to keep firmly in mind that all their evidence is necessarily hypothetical and that hypothetical evidence by its very nature lends itself to exaggeration and embellishment in the interests of the party on whose behalf it is given. It is very easy for an underwriter to convince himself that he would have declined a risk or imposed special terms if given certain information. For this reason, such evidence has to be rigorously tested by reference to logical self-consistency, and to such independent evidence as may be available.

29 Often referred to as the "minimum performance principle": see *Lavarack v Woods of Colchester Ltd* [1967] 1 QB 278 (CA).
30 *Camarata Property Inc. v Credit Suisse Securities (Europe) Ltd* [2011] EWHC 479 (Comm). This is not a case about construction professionals but the approach taken to professional conduct is of general application.
31 [1995] 1 WLR 1602 (CA). For an application in the construction context see *Trustees of Ampleforth Abbey Trust v Turner & Townsend Project Management Ltd* [2012] EWHC 2137 (TCC). One of the allegations against the project manager was that it failed to take adequate steps to persuade the contractor to execute the contract. The assessment of loss involved a consideration of what the contractor would have done had those steps been taken. In particular, the court had to assess the chance that the contractor would have agreed to certain terms.

action of the third party is dependent upon some hypothetical act of the claimant the claimant must show on the balance of probability that he would have acted in that way; (ii) the Court then assesses the chance that the third party would have acted as the claimant contends he would have; in rare cases evidence may be available from the third party to assist the Court; in the usual case the Court will have to do its best on the basis of the information available; (iii) provided that the Court is satisfied that there is a "substantial chance"[32] that the third party would so have acted (being a chance which is more than merely fanciful) the Court applies that chance (as a percentage) to the loss that the claimant can show results if the third party had acted as contended for.

6.42 In *Trustees of Ampleforth Abbey Trust v Turner & Townsend Management Ltd*,[33] the claimant contended that because of breaches of duty by its project managers it lost the chance to persuade its building contractor, which had been proceeding on the basis of letters of intent, to enter into a building contract. The court found that there had been a breach of contract. The question was what loss was suffered? Following *Allied Maples*, the Judge ascertained the benefit to the claimant had a building contract been entered into (which involved estimating the increase in the value of the claim which the claimant would have had against the contractor) and then reduced that sum by one third to take account of the fact that there was a two thirds chance that the contractor would have entered into the contract.

SECTION E: ASSESSMENT OF LOSS

Costs of rectification or diminution in value?

6.43 The object of an award of damages is to put the innocent claimant "in the same position as he would have been in if he had not sustained the wrong for which he is now getting compensation or reparation".[34] This can be done either by compensating the claimant for what it has lost, or by indemnifying the claimant against the consequences of undertaking the work to be restored to the position it would have been in but for the breach. Which of these two alternatives is adopted will depend upon the facts of the case. Moreover, depending on the facts elements of both cost of repair and indemnification may be appropriate and/or there may be other heads of loss which are properly recoverable.

6.44 In an action against a contractor or a professional for defective work, the appropriate measure of loss has usually been taken to be the cost of reinstatement/repair, because that was the foreseeable consequence of the defective work[35] The repair or reinstatement costs will be those appropriate to the quality of the building or other structure which the employer was entitled to expect.[36]

6.45 However, the general rule that a claimant is entitled to the costs of reinstatement is subject to exceptions.

---

32 Something which courts in other fields have put at a chance exceeding 10–15%.
33 [2012] EWHC 2137 (TCC).
34 *Livingstone v Rawyards Coal Company* (1880) 5 App. Cas. 25 at 39.
35 Per HH Judge Coulson QC in *McGlinn v Waltham Contractors* in [2007] EWHC 149 (TCC).
36 *Catlin Estates Ltd v Carter Jonas* [2005] EWHC 2315 (TCC).

6.46 If the claimant only has a limited interest in the property,[37] or if he could obtain a satisfactory replacement for the property by buying elsewhere, then it would not be foreseeable that he would carry out repair/reinstatement, and his loss would be accurately assessed by reference to the diminution in the value of the land or the cost of purchasing a replacement.[38]

6.47 Where the costs of rectification are disproportionate to the benefit which the claimant will obtain by rectification the court may choose to award diminution in value.[39]

6.48 Conversely where the costs of rectification would not fully compensate the claimant the court will award damages based on some other measure, often diminution in value.[40] It *may* be necessary to look to the expenditure which the claimant has "wasted" in a situation where, had competent advice been given, the employer would not have proceeded at all.[41]

6.49 It may be appropriate to award both cost of repair and diminution in value if the fact of the defect has caused both a need for repair and some other economic loss[42] (for example defects leading to underpinning and resultant stigma, or defects leading to the inability to obtain certificates or guarantees).

6.50 It is the nature of the compensatory exercise that the claimant should not be put in a better position than it would have been had the breach of duty not taken place.[43] It follows, therefore, that if the construction professional's negligence consists in not advising that a more robust foundation design be utilised, the claimant must give credit for the additional costs it would have had to pay to the contractor for the additional cost of the larger foundations. If the construction professional's breach of duty consists in failing to advise the employer to obtain a particular survey the claimant's claim must give credit for the costs that would have been incurred. In the hypothetical situation where the "competent" advice is given the construction professional itself may have been entitled to further payments.

6.51 One seeming exception to this "no overcompensation" principle is where the rectification of the defect necessarily confers an additional benefit upon the claimant (for example, because changes in building regulations require that the structure be constructed to a

---

37 See *Saigol v Cranley Mansions Ltd (No. 2)* (2000) 72 Con LR 54 (CA).
38 Per HH Judge Coulson QC in *McGlinn v Waltham Contractors Ltd* [2007] EWHC 149 (TCC).
39 *Ruxley Electronics and Construction Ltd v Forsyth* [1996] AC 344 (HL). The swimming pool was 6 ft deep rather than 7 ft 6 inches deep as contracted for. As constructed it was perfectly usable. The House of Lords awarded compensation for loss of amenity, there being no diminution in value.
40 In *Saigol v Cranley Mansions Ltd (No. 2)* (2000) 72 Con LR 54, negligent design and supervision by a building surveyor led to the claimant losing her home, which was repossessed. The Court of Appeal rejected the contention that the claimant's claim was confined to the cost of putting right the defects. She was awarded the diminution in value of the flat as a result of the defects.
41 This suggestion was made by Ramsey J in *Cooperative Group Ltd v John Allen Associates Ltd* [2010] EWHC 2300 (TCC) at [339]. It is suggested that the measure may be difficult to apply in cases where he claimant has nonetheless received some benefit.
42 See *Hoadley v Edwards* [2001] PNLR 41 (Ch).
43 Suppose an architect negligently fails to spot defects when inspecting the contractor's work. The appropriate loss awarded against the architect as damages may be the cost of remedying the defects but only if the claimant can show that, had the architect acted competently, the contractor would have rectified the work at no cost to the claimant. If this would not have happened (for example, because the contractor was insolvent) there may be no loss. If it *might* have happened (for example because the contractor was in dispute with the employer and subsequently became insolvent) the claimant's loss may be a loss of a chance, the chance being applied to the costs of rectification.

higher standard than had been provided for in the contract). It is suggested that the correct analysis of this problem is provided by His Honour Judge Newey QC in *Richard Roberts Holdings Ltd v Douglas Smith Stimson Partnership*:[44]

> If the only practicable method of overcoming the consequences of a defendant's breach of contract is to build to a higher standard than the contract had required, the plaintiff may recover the cost of building to that higher standard. If, however, a plaintiff needing to carry out works because of a defendant's breach of contract, chooses to build to a higher standard than is strictly necessary, the courts will, unless the new works are so different as to break the chain of causation, award him the cost of the works less a credit to the defendant in respect of betterment.

Other expenditure or costs

6.52 Breach of duty by a construction professional may lead to wasted costs, particularly if a project is aborted or part of a project has to be undertaken again. In *Riva Properties Ltd and ors v Foster and Partners Ltd*,[45] the architect failed to advise the claimant that the hotel which the claimant believed could be completed for a total outlay of £100 million could not in fact be completed for anything like that sum. The work was abandoned and, amongst other claims, the claimant contended that it was entitled to the wasted costs of time spent by the architect and others on the project. The court agreed. Because the work needed to be done again, the best way of assessing the claimant's loss was by reference to the value of the wasted costs.

6.53 The employer may have overpaid the contractor or some other person (for example, as a result of negligent certification). The claimant will often suffer additional costs which fall outside the costs or reasonable repair and reinstatement. These might include costs of finance if the claimant has borrowed to support the project, costs of alternative accommodation if the claimant needs to occupy other premises until the repairs are carried out, or costs representing loss of benefit from the property.

6.54 This last head of loss may be problematic both in terms of remoteness and in terms of evaluation. It is relatively straightforward to ascertain a construction professional's liability in circumstances where it has caused an employer to lose six months' rent in respect of a development. It is much more difficult to ascertain that liability if the employer did not intend to make a commercial use of the premises.[46]

6.55 Financial losses of various different kinds may be recoverable against a construction professional either as the loss directly caused by the breach of duty, or as a foreseeable consequence of directly caused loss.

6.56 Directly caused financial losses are less common. These can arise in situations where the breach of duty by the construction professional causes the client to lose some financial advantage or to incur some financial burden. The architect whose breach of duty causes the building development to be delayed may be liable to the client for the financial consequences of the delay, including (in an appropriate case) loss of profits. The client who decides to purchase a derelict property on the basis of advice from the quantity surveyor

---

44 (1988) 46 BLR 50 at 69.
45 [2017] EWHC 2574 (TCC).
46 In *Bella Casa Ltd v Vinestone* [2005] EWHC 2807 (TCC), the owner of property failed to obtain general damages for being unable to use it.

as to the costs of a particular refurbishment scheme *may* be able to recover the difference between what it spent and what the property was really worth once it was appreciated that the cost of refurbishment was in fact far higher.

6.57 Financial losses flowing from damaged or defective works are more usual. The architect whose design of fire-stopping was inadequate may be liable not just for the costs of repair to a fire damaged building, but also for the loss of profits for the period when the building was unusable. The engineer whose foundations design leads to the supermarket floor deflecting will be liable for the costs of remedial work but very likely also the loss of profit for the period during which the supermarket is closed whilst those remedial costs take place. It is not uncommon in such cases for the financial costs to be much greater than the costs of repair themselves.

Liability to third parties

6.58 Breach of duty by a construction professional may lead to some other party (usually the client) becoming exposed to a claim by some other person. For example, a design error by the services engineer which leads to a fire which spreads to neighbouring premises may have the result that the employer is exposed to a claim for fire damage and its consequences from the damaged neighbouring properties. Negligent design by the architect, causing delay to a project, may have the consequence that the design and build contractor (to whom the architect's appointment has been novated) becomes exposed to a claim for liquidated damages by the employer. Negligent contract administration may lead to the employer becoming exposed to a claim for loss and expense by the contractor.

6.59 Claims of this nature do not have to be proven before the construction professional can become liable. Indeed, they do not even have to be liquidated (that is, costed). It is sufficient to ground a cause of action that the construction professional's breach of duty has caused the client (or some other party to whom a relevant duty is owed) to become exposed to a claim.

6.60 Where a claimant has settled a claim with another party and then seeks to assert that its loss representing that settlement was caused by a construction professional it is not always obvious that the construction professional has caused the loss even if the construction professional has acted in breach of duty. A complex investigation is merited. In *Siemens Building Technologies FE Ltd v Supershield Ltd*,[47] Ramsey J explained the approach which should be taken:

> *(1) For C to be liable to A in respect of A's liability to B which was the subject of* a settlement it is not necessary for A to prove on the balance of probabilities that A was or would have been liable to B or that A was or would have been liable for the amount of the settlement. (2) For C to be liable to A in respect of the settlement, A must show that the specified eventuality (in the case of an indemnity given by C to A) or the breach of contract (in the case of a breach of contract between C and A) has caused the loss incurred in satisfying the settlement in the manner set out in the indemnity or as required for causation of damages and that the loss was within the loss covered by the indemnity or the damages were not too remote. (3) Unless the claim is of sufficient strength reasonably to justify a settlement and the amount paid in settlement is reasonable having regard to the strength of the claim, it cannot be shown that the loss has been caused by the relevant eventuality or breach of contract. In assessing the strength of the claim, unless the

---

47 [2009] EWHC 927 (TCC).

claim is so weak that no reasonable party would take it sufficiently seriously to negotiate any settlement involving payment, it cannot be said that the loss attributable to a reasonable settlement was not caused by the eventuality or the breach. (4) In general if, when a party is in breach of contract, a claim by a third party is in the reasonable contemplation of the parties as a probable result of the breach, then it will generally also be in the reasonable contemplation of the parties that there might be a reasonable settlement of any such claim by the other party. (5) The test of whether the amount paid in settlement was reasonable is whether the settlement was, in all the circumstances, within the range of settlements which reasonable people in the position of the settling party might have made. Such circumstances will generally include: (a) The strength of the claim; (b) Whether the settlement was the result of legal advice; (c) The uncertainties and expenses of litigation; (d) The benefits of settling the case rather than disputing it. (6) The question of whether a settlement was reasonable is to be assessed at the date of the settlement when necessarily the issues between A and B remained unresolved.

6.61 As is discussed elsewhere,[48] it is not uncommon for construction professionals to be joined to court proceedings taken against the employer on the basis that if the employer is liable to the claimant then that liability was brought about by a breach of duty on the part of the construction professional. In some cases a client will proceed against the construction professional in advance of the determination of the claim against it from the third party, seeking declaratory relief that it is entitled to be indemnified against any loss it may sustain in respect of that claim.

Personal injury, inconvenience and distress

6.62 Just like any other person causing personal injury a construction professional can be liable to the victim of its breach of tortious duty if that person suffers personal injury.[49] In the same way the construction professional may be liable to the dependents of any person killed as a result of its breach of duty. As set out above, the construction professional's breach of duty may lead to some other person (usually the client) becoming liable for damages for personal injury or death.

6.63 Such cases are rare and the complexities of awards of damages for these heads of loss are outside the scope of this work. However, it may be a relevant consideration for any construction professional in a modest way of business which is deciding on its level of professional indemnity cover that personal injury damages for the most serious cases frequently run into the millions and may comfortably exceed the values of the modest properties in respect of which the professional is generally engaged.

6.64 Like all professional persons a construction professional *may* be liable to its client[50] for inconvenience, distress and loss of amenity. These are contractual claims. Their underlying justification is that it is part of the construction professional's bargain with its client that it will provide peace of mind, or pleasure, or a measure of comfort.[51] It follows that the scope of such claims will be limited.

---

48 See Chapter 9.
49 Plainly the victim could be the client – for example where the architect designs a residential building.
50 Or conceivably a third party closely related to the client and for whose benefit the construction professional is engaged – see *West v Ian Finlay and Associates* [2014] EWCA Civ 316.
51 *Farley v Skinner (No. 2)* [2002] 2 AC 732 (HL) is an example of a contract where part of the bargain was that the valuer would provide the client with peace of mind.

CAUSATION LOSS AND DAMAGE IN CLAIMS AGAINST CONSTRUCTION PROFESSIONALS

6.65 Moreover, the designer of residential property may more readily be said to take on such a responsibility than the contract administrator engaged on the same project. An individual whose house is draughty and prone to leaks may suffer both inconvenience and distress, the avoidance of which was an important part of the bargain with the design professional. A limited company may have responsibilities to persons who suffer inconvenience and distress because they use its draughty and leaky building, but by definition it cannot suffer that kind of damage itself. Loss of amenity – the provision of something which does not need repair but which, often for aesthetic reasons, is less valuable to the client than it should be – is necessarily difficult to gauge and likely to be limited to unusual facts.[52] Awards in this area tend to be very modest, often £2,000 or £3,000.[53]

The construction professional's fees

6.66 As set out above, the damage suffered by a claimant who has engaged a construction professional may include wasted fees paid to that construction professional. However, claimants sometimes go further in attacking the construction professional's entitlement to fees and argue that payment of fees can be refused on the grounds of the construction professional's breach of duty, even though these fees may not have been wasted.

6.67 Thus take the situation of an architect engaged on a modest residential project. The project proceeds to a stage where the building is two-thirds complete. At this stage the architect is owed £20,000. The architect then omits to serve a valid party wall notice with the effect that the works are halted. The employer dismisses the architect but is fortunate enough to engage a replacement at minimal additional cost. The completion is not in fact delayed. The dismissed architect requires payment of his fees.

6.68 Can the employer refuse to pay these on the basis that the architect did not complete the works it was engaged to undertake? The answer is almost certainly "no". In the main this is because it would be an unusual engagement which did not contain a provision that the construction professional was to be paid in stages. It is suggested that it would require very unusual facts for a construction professional's contract to be viewed as requiring it to complete all work before being entitled to anything.[54]

6.69 In theory, a construction professional's services could be of so little value that there has been a "total failure of consideration" (the client has received absolutely nothing of value). In those circumstances the contract is voidable and the client may be entitled to recover any remuneration paid and to refuse to pay any further remuneration. Save where the construction professional has only had the briefest and most disastrous involvement on the project, it is unlikely that such arguments will succeed.[55]

---

52 A small award of damages was made in *Ruxley Electronics and Construction Ltd v Forsyth* [1996] AC 344 (HL) for a swimming pool which was shallower than the client wanted, but perfectly adequate for its intended purpose.
53 In *AXA Insurance UK plc v Cunningham Lindsey United Kingdom* [2007] EWHC 3023 (TCC), the court indicated that it would a rare case in which damages for distress and inconvenience would exceed £2,500.
54 See *Multiplex Constructions (UK) Ltd v Cleveland Bridge UK Ltd* [2006] EWHC 1341 (TCC) and *William Clark Partnership Ltd v Dock St PCT Ltd* [2015] EWHC 2923.
55 For a case where that argument nearly succeeded, see *Riva Properties Ltd and ors v Foster and Partners Ltd* [2017] EWHC 2574 (TCC).

SECTION F: MITIGATION

6.70 Properly analysed, "mitigation" is an umbrella term which encompasses a number of related approaches to the assessment of damages. Mitigation is concerned with actions taken by the claimant, in the knowledge of the breach,[56] which either makes it unjust for the claimant to recover all of the loss it has sustained or, by the operation of the same policy, makes it just that the claimant recovers additional loss which it has incurred in attempting to remediate the consequences of the breach.

6.71 The first sense is that most generally encountered: the claimant may not recover loss which it should have prevented by acting reasonably.[57] This is sometimes spoken of as a "duty":[58]

> The plaintiff has, whether as part of the requirement that he act reasonably or otherwise, a duty to mitigate his loss. This may require him if presented with two or more choices to choose the one which will keep his losses to the minimum. If he is incurring loss because he cannot use his property, his duty to mitigate may require him to repair it as quickly as possible, even if early repairs would cost more than later repairs would. The duty to mitigate may require the plaintiff to have regard to advice from third parties, or even from the defendant, or from the defendant's advisers.[59]

6.72 If there are two equally efficacious alternative remedial schemes, and one is cheaper than the other, then *prima facie* the claimant is obliged to put in hand the cheaper of the two schemes.

6.73 A claimant must act reasonably but the standard of reasonableness is generous having regard to the predicament in which it finds itself.[60] Thus it will be a rare case in which a claimant which has relied upon (seemingly) competent advice can be said to have acted unreasonably, such advice being a "highly significant factor".[61]

6.74 Each case will turn upon its facts:

> The question of whether advice of an expert, even if professionally reasonable, can convert expenditure into reasonable expenditure involves a consideration of the facts in any given case. There must be some effective causal link between the incurrence of the expenditure upon the advice of the expert and the breach of contract. Thus, where a garden wall has been constructed defectively by a builder, the reconstruction of that wall may well be the basic measure

---

56 It is this knowledge which takes mitigation outside the strict parameters of the break in the chain of causation cases.

57 *British Westinghouse Electric & Manufacturing Co Ltd v Underground Electric Railways Co of London Ltd (No. 2)* [1912] AC 673 (HL) at 689.

58 "Duty" here is a shorthand for a requirement which is a precondition to recovery. There is no legal duty to mitigate: *Darbishire v Warran* [1963] 1 WLR 1067 (CA).

59 Per His Honour Judge Newey QC in *Board of Governors of the Hospitals for Sick Children v McLaughlin & Harvey plc and ors* (1987) 19 Con LR 25 at 96.

60 In *Banco De Portugal v Waterlow & Sons Ltd* [1932] AC 452 (HL), Lord Macmillan said, at p. 506:

> Where the sufferer from a breach of contact finds himself in consequence of that breach placed in the position of embarrassment the measures which he may be driven to adopt in order to extricate himself ought not be weighed in nice scales at the instance of the party whose breach of contract has occasioned the difficulty. It is often easy after an emergency has passed to criticise the steps which have been taken to meet it, but such criticism does not come well from those who themselves created the emergency. The law is satisfied if the party placed in a difficult situation by reason of the breach of a duty owed to him has acted reasonably in the adoption of remedial measures, and he will not be held disentitled to recover the costs of such measures merely because the party in breach can suggest that other measures less burdensome to him might have been taken.

61 Per Waller LJ in *Skandia Property UK Ltd v Thames Water Utilities Ltd* [1999] BLR 338 (CA).

of damages; if, during the construction process, the client's surveyor or engineer recommends, perfectly reasonably, that a wall of the house should be repointed, that does not mean that the cost of the repointing will be recoverable as damages. If the advice of the expert is merely tangential or coincidental to the work the cost of which is recoverable as damages, the costs of the work carried out to that extent upon the expert's advice will, generally, not be recoverable.[62]

SECTION G: LIMITATION OF ACTIONS

6.75 It is a commonplace of modern civil law jurisdictions that some time limit is placed upon claimants who wish to bring actions to recover damages or obtain some other remedy.

6.76 The system of time limits in English law is applied by statute (the Limitation Acts) and it generally operates to restrict the ability of a claimant to achieve a remedy against a defendant by requiring that proceedings[63] be commenced within a certain period of the cause of action arising. The law of limitation of actions is complex and the reader should refer to the analysis in the specialist texts.[64] What follows is a summary of the law as it usually applies to construction professionals.

6.77 The general scheme of limitation in English law is as follows:

- claims to recover damages in contract will generally become "statute barred" if the proceedings are commenced more than six years after the date of the breach of contract;[65]
- claims to recover damages in tort (for example negligence) will generally become "statute barred" if the proceedings are commenced more than six years after the date of the first damage;[66]
- the first exception to the general rule is that it is open to contracting parties to agree a longer period[67] for claims to be brought should they so wish; this is generally done by the parties entering into a deed (under which the limitation period is 12 years);[68] but it can also be done by the terms of the contract if they are sufficiently clear;[69]
- the second exception is that Parliament allow a special extension of the period for bringing negligence claims in situations where the claimant does not learn

---

[62] Akenhead J in *Axa Insurance UK plc v Cunningham Lindsay United Kingdom* [2007] EWHC 3023 (TCC) at [267].

[63] That means any kind of legal claim, whether brought in adjudication, arbitration or in court.

[64] For example, McGee, *Limitation Periods* (Seventh Edition).

[65] Limitation Act 1980, section 5.

[66] Limitation Act 1980, section 2.

[67] It is possible for the parties to agree a *shorter* period after which claims are barred. The words used to effect this restriction must generally be clear as the courts do not readily accept that parties give up important rights against each other. See, for example, *Oxford Architects Partnership v Cheltenham Ladies College* [2006] EWHC 3156 (TCC).

[68] Section 8 Limitation Act 1980 – a speciality is a contract under seal.

[69] One method of doing this (admittedly rare in the engagement of construction professionals) is to require the construction professional to indemnify the client for any loss that it suffers by reason of particular kinds of breaches by the construction professional. An agreement to indemnify often operates to make the indemnifier liable at the point at which the person to be indemnified becomes liable for an ascertained sum of money. The most common contractual extension of the limitation period occurs long after the contract has been agreed. It is very common for claimants and those acting for construction professionals to enter into "standstill agreements" where the operation of the Limitation Acts is suspended, see Chapter 9.

about the damage for some time; the Limitation Acts provide for an alternative limitation period in negligence whereby claims are barred if commenced more than three years after the claimant had "knowledge" of the material facts about the damage.[70]

6.78 The contracts of many construction professionals are in the form or deeds precisely so that the employer can take advantage of the 12-year limitation period. In such situations is common for collateral warranties given to third parties to stipulate that they may commence proceedings against the construction professional at any time within 12 years of practical completion. The Limitation Acts do not operate to shorten these contractual agreements.

6.79 Outside of cases where the parties have agreed to extend the limitation period, the tortious obligations of the construction professional become important. For the reasons which are given above, construction professionals will always owe their clients a tortious duty of care which is co-extensive with their contractual duties. However, the limitation period in respect of negligence claims is more generous than the contractual limitation period of six years from the date of breach of contract. It is more generous in two different ways, both of which merit some explanation.

6.80 Whilst the claimant often suffers the first "damage" flowing from a breach of contract at the moment the breach of contract occurs, this is not always the case. Under English law, "damage" is not equated with breach: the fact that the claimant has received a lesser service than it contracted for is not necessarily the same thing as saying that it has suffered "damage". "Damage" has been defined by the courts in terms of physical damage or economic harm.[71] To take a simple example, the architect may produce a defective drawing; the drawing may be handed to the contractor; the contractor may act upon it. "Damage" probably only occurs in the last of these stages. Until that point the claimant cannot point to any act or omission on the part of the architect which has caused it loss. Thus if the architect corrects the drawing before the contractor acts on it and the contractor makes no claim in respect of that correction, there is no "damage".

6.81 The point at which physical damage occurs is when the first manifestation of the defective nature of design occurs, as matter of objective fact, to be ascertained on the basis of the technical evidence.

6.82 In *Pirelli General Cable Works Ltd v Oscar Faber & Partners*,[72] engineers were engaged to design a factory including its chimney. The chimney was constructed by the end of 1969. Unbeknown to anyone at the time, cracking began in the chimney in April 1970. The cracking would have been observable for the first time in October 1972. The employer did not in fact discover the cracks until November 1977. Proceedings were commenced in October 1978. This was before the introduction of what is now Section 14A of the Limitation Act 1980, which allows a more generous limitation period for hidden defects. Consequently the question for the House of Lords was, when did damage occur? The House of Lords rejected the suggestion that there was damage when the chimney was constructed

---

70 Limitation Act 1980, section 14A.
71 See *Nykredit Mortgage Bank Plc v Edward Erdman Group Ltd (No. 2)* [1997] 1 WLR 1627 (HL) at 1630B-1630G.
72 [1983] 2 AC 1 (HL).

(even though its design was defective), and rejected the suggestion that damage occurred when anyone wanting to buy the factory would have paid less for it because of observable cracking in the chimney.

6.83 The *Pirelli* test is open to criticism because there is no very satisfactory distinction between the concept of a defective building (a chimney which is likely to crack) and a damaged building (a chimney which has cracked). This is particularly the case because the House of Lords allowed an exception for buildings which were so badly designed they were doomed from the start.[73] *Pirelli* is also inconsistent with the approach taken in other comparable jurisdictions.[74] However, it remains the authoritative statement of the law in England and Wales.[75] Given the infrequency with which *Pirelli* type facts are likely to matter now that Section 14A of the Limitation Act 1980 provides a limitation period tailored to hidden defects, it may remain the law for some time.

6.84 Damage in the form of purely economic harm, such as the exposure of the client to a claim by a third party, almost always occurs at the moment when the act or omission has the effect of making the client so exposed. It does not occur when the claim is made. This is because "damage" is generally accepted to include "any detriment, liability or loss capable of assessment in money terms and it includes liabilities which may arise on a contingency over which the plaintiff has no control".[76]

6.85 The courts have not always been willing to accept that exposure to contingent claims occurs at the moment when the construction professional's actions or omissions lead the possibility of exposure. In *Linklaters Business Services v Sir Robert McAlpine Ltd*,[77] Akenhead J held that the construction professional's liability to its client in respect of the contingent claim did not arise until the claim was made. This decision probably turns on its special facts. The better view, it is suggested, is that absent unusual facts, a contingent liability becomes damage when it is created and not when the contingency comes to fruition.[78]

6.86 For construction professionals the most important limitation provision is Section 14A of the Limitation Act 1980. This section provides a three year limitation period where time starts to run from:

> the earliest date on which the plaintiff or any person in whom the cause of action was vested before him first had both the knowledge required for bringing an action for damages in respect of the relevant damage and a right to bring such an action.[79]

6.87 The section is complex and much litigated. It is beyond the scope of this book to provide a comprehensive analysis. In essence, it works in this way:

- "Knowledge" is knowledge about the material facts, which means such facts as would lead a reasonable person to consider the damage sufficiently serious to

---

[73] For an example of a "doomed from the start" case see *New Islington and Hackney Housing Association v Pollard Thomas Edwards Ltd* [2001] PNLR 515.

[74] See *Invercargill City Council v Hamlin* [1996] AC 624 (PC), a decision of the Privy Council concerning the law of New Zealand, and *Bank of East Asia Ltd v Tsien Wui Marble Factory Ltd* [2001] 1 HKLRD 268, concerning the law in Hong Kong.

[75] *Abbott v Will Gannon & Smith Ltd* [2005] EWCA Civ 198.

[76] *Nykredit Mortgage Bank Plc v Edward Erdman Group Ltd (No. 2)* [1997] 1 WLR 1627 (HL) at 1630B-1630G.

[77] [2010] EWHC 2931 (TCC).

[78] See *Co-operative Group Ltd v Birse Developments (in liquidation)* [2014] EWHC 530 (TCC).

[79] Section 14A(5).

justify starting proceedings against a defendant who did not dispute liability and was able to satisfy a judgment[80] (in other words, damage beyond that which is trivial).
- "Knowledge" must be knowledge that the damage was attributable to the actions or omissions of the construction professional although the claimant does not have to know that, as a matter of law, the construction professional's acts or omissions amounted to negligence[81] (in other words, the claimant has to know that the construction professional has caused the problem although it may not know that it has a good case against the construction professional).
- "Knowledge" includes knowledge which the claimant should have obtained acting reasonably (both from its own appreciation of the facts and from any expert evidence which it might have been expected to obtain[82] (the lazy or indifferent claimant will not be put in a better position than the hypothetically careful and conscientious claimant).

6.88 The section applies to specific causes of action. A claimant may only have one cause of action against the geotechnical engineer which produces a report containing a number of errors: the breach of duty consists in producing a defective report. If the claimant has knowledge in respect of just one of these errors more than three years prior to the commencement of proceedings, the entire claim may be statute barred.[83] By contrast, a claimant may have multiple causes of action against a contract administrator, which may be alleged to have committed a number of separate breaches of duty. In the latter situation, some causes of action may be statute barred because the claimant had knowledge more than three years prior to the commencement of proceedings and some may not.[84]

6.89 It will be clear that the operation of the section is highly fact sensitive. The reported cases are no more than examples of the application of the provision to particular facts.[85]

SECTION H: CONTRIBUTORY NEGLIGENCE

6.90 In many claims against construction professionals the claimant will have acted in a way which may have contributed to the loss that it has suffered, either by contributing to the cause of the loss or by contributing the extent of the loss. Sometimes those acts or omissions will be careless in and in those circumstances a defendant construction professional may be entitled to ask the court or arbitrator to reduce any award of damages against it so as to take into account the contribution made by the claimant.

6.91 The jurisdiction for this power is the Law Reform (Contributory Negligence) Act 1945 which provides that where a person suffers damage:

---

80 Section 14A(7).
81 Section 14A(8) as read with 14A(9).
82 Section 14(10). If expert evidence is obtained and fails to provide knowledge the claimant is not deemed none the less to have knowledge: although it is deemed to act reasonably it is not deemed to have engaged a reasonable expert.
83 *Eagle v Redlime Ltd* [2011] 136 Con LR 137.
84 See *Hunt v Optima (Cambridge) Ltd* [2015] 1 WLR 1346.
85 For an example concerning architects and the claimant's "knowledge" in respect of various defects see *Renwick v Simon and Michael Brooke Architects* [2011] EWHC 874 (TCC).

as a result partly of his own fault and partly of the fault of any other person or persons ... the damages recoverable in respect thereof shall be reduced to such extent as the court thinks just and equitable having regard to the claimant's share in the responsibility for the damage.[86]

"Fault" here is equated with negligence or carelessness.

6.92 There may be some rare cases where contributory negligence is not available as a matter of law because the construction professional is in breach of an absolute contractual obligation, rather than an obligation to use reasonable skill and care,[87] but for all practical purposes it can be assumed that a contributory negligence defence will always be available where the facts will sustain it.

6.93 However, the courts are not generous in finding situations where the facts will sustain a reduction of the construction professionals' damages because of the contributory negligence of the claimant.

6.94 The burden of proof lies with the construction professional. It must establish that the claimant was careless (amounting to a negligent disregard of its own interests), that its negligence caused loss and that this is the same loss which the construction professional has caused.

6.95 Particularly in large construction projects, the claimant will act through and on the advice of a number of different agents and it will be rare case in which the court will attribute the negligence of these agents to the client.[88] Thus the failure to deal with the presence of unexpected structures in the ground may have been the shared responsibility of the architect and the contract administrator, but the court is unlikely to reduce the employer's damages from the architect for that reason.

6.96 The courts are also reluctant to make substantial reductions to damages for alleged negligence which goes to the very task which the construction professional is meant to perform. At its simplest, this approach to fault makes it difficult, if not impossible, for the construction professional to contend that because the client was itself experienced and knowledgeable in construction it should have spotted the fact that construction professional had made a mistake. At a more refined level it finds expression in the unwillingness of the courts to make substantial awards to claimants for causing losses where it was the obligation of the construction professional to protect the client against precisely that eventuality.

6.97 In *Pride Valley Foods Ltd v Hall & Partners (Contract Management) Ltd (No. 1)*,[89] the project managers were liable to the employer for failing to warn that panels were highly combustible. A fire occurred through the negligence of the employer and spread rapidly because of the panels. The Court of Appeal indicated that that the award of 50% contributory negligence appeared high.[90] Sedley LJ said that no deduction for contributory negligence should be made where the harm caused by the claimant's own negligence was wholly within the very risk which it was the defendant's duty to guard against. In *Sahib Foods Ltd (in liquidation) v Paskin Kyriades Sands*[91] a different Court of Appeal thought that this went too far, but agreed that if part of the defendant's duty was to protect

---

86 Section 1(1).
87 See *Forsikringsaktieselskapet Vesta v Butcher* [1989] AC 852 (HL).
88 The construction professional's remedy is to seek a contribution from the agent – see below.
89 (2001) 76 Con LR 1.
90 Although they declined to interfere, the amount being a matter for the Judge's discretion.
91 [2003] EWHC 142 (TCC).

the claimant against the consequences of its own negligence this should be taken into account.[92]

6.98 The overall assessment of what is "just and equitable" takes into account questions of causative potency and "blameworthiness". Reported decisions in this area very much turn on their own facts and very much reflect an impressionistic approach to what is "just and equitable".[93]

SECTION I: CONTRIBUTION

6.99 "Contribution" is the way in which a person liable to someone else can obtain a remedy from a third person who is also liable for the same damage. It is very important in actions involving construction professionals because it is common for losses sustained by employers to have been caused by two or more persons.

6.100 The operation of the remedy can be illustrated by a simple example. A construction project is delayed because there is a design clash between the architectural design and the services design. The architect blames the services engineer and vice versa. The employer decides that blame chiefly lies with the architect and it claims against the architect for all of the losses caused by the delay. The architect defends that claim, contending that it acted competently and that the fault lies with the services engineer. However, it also issues contribution proceedings against the services engineer seeking a contribution in the event that its defence against the employer fails. The matter proceeds to trial. The architect is found to be liable to the employer for the whole loss but the court also finds that the services engineer was 30% to blame. It awards the architect a contribution of 30% of the damages and costs that it must pay to the employer.

6.101 The right is a creature of statute. Section 1(1) of the Civil Liability (Contribution) Act 1978 provides:

> Entitlement to contribution.
>
> (1) Subject to the following provisions of this section, any person liable in respect of any damage suffered by another person may recover contribution from any other person liable in respect of the same damage (whether jointly with him or otherwise).

6.102 There are a number of important ingredients of this provision (as provided by subsequent sub-sections). Both the person seeking contribution and the person from whom contribution is sought must be liable to a third person.

- Liability means any liability that could be established in England and Wales.[94]
- The person seeking contribution is "liable" even though it may have settled the claim against the person to whom it was liable provided the settlement was bona fide.[95]

---

92 On the unusual facts in that case the Court of Appeal thought that a reduction of two thirds was appropriate.

93 See for example *Lloyds Bank plc v McBains Cooper Consulting* [2015] EWHC 2372 (TCC), where the Bank's damages as against the project monitor were reduced by only 30% notwithstanding its carelessness in making the loan.

94 Section 1(6).

95 Section 1(4). The point of this provision is to prevent unnecessary litigation and to encourage settlement. A construction professional which is liable to an employer should not have to obtain a court order to that effect

- The person from whom contribution is sought is "liable" to the third person even if (at the moment contribution is sought) he had ceased to be liable because of limitation.[96]
- The liability of the person seeking contribution and the liability of the person from whom contribution is sought must be for the "same damage".

6.103 The requirement that both the person seeking contribution and the person from whom it is sought must be liable for the same damage has given rise to a number of disputes. The issue of "same damage" tends to be very straightforward where two construction professionals cause the same defect or the same delay. It becomes more complicated where a construction professional and another person cause harm or loss which can be described in different ways and which involves different kinds of contractual or tortious rights.

6.104 In *Royal Brompton Hospital NHS Trust v Hammond (No. 3)*,[97] the employer contracted with the contractor and engaged the architect as, amongst other things, contract administrator.[98] The project was delayed and the architect certified extensions of time. This had the consequence of entitling the contractor to loss and expense and disentitling the employer from levying liquidated damages. In a subsequent arbitration the employer settled making a payment to the contractor. The employer then sued the architect for negligent certification. The architect sought contribution from the contractor, contending that it was liable to the employer for the same damage. It said that the damage was the lost payment of loss and expense and the lost liquidated damages. However, the House of Lords ruled against the architect. "Same damage" meant the same harm – it did not mean the same damages or similar harm. The essence of the contractor's liability to the employer was the delay in completion. The essence of the architect's liability to the employer was an impairment of its rights against the contractor. These were different kinds of damage.

6.105 It is suggested that "same damage" problems will be relatively rare in disputes involving construction professionals. Where the damage consists of a defect which is caused by A and missed by B (carrying out its inspection function) both A and B will be liable for the same damage.[99] Where both A and B cause damage to the employer but the damage caused by B is a small part of the damage caused by A, they will still be liable for the same damage.[100]

---

before it can claim contribution. Indeed, provided that the claim as made out against that person would have resulted in liability if the facts asserted were true, the person from whom contribution is sought cannot be heard to say that there was no liability in the first place: *IMI plc v Delta Ltd* [2016] EWCA Civ 773.

96 Section 1(3). One exception to this provision is where a contractual agreement between the person from whom contribution is sought and the person to whom it is said to be liable removes the liability. In *Bloomberg LP v Sandberg (a firm)* [2015] EWHC 2858 (TCC), a construction professional sought contribution from a design and build contractor. The contractor contended that the terms of its collateral warranty prevented the beneficiary from suing after 12 years and that this removed the right of action. The contention failed because the contractual provision was held to be procedural bar: it prevented recovery (as does limitation) but it did not rule out liability.

97 [2002] UKHL 14.

98 Strictly speaking, "project architect" under the relevant contract.

99 *McConnell v Lynch-Robinson* [1957] NI 70, a case involving a contractor and an architect, cited in *Royal Brompton*.

100 *Bank of Ireland v Faithful & Gould Ltd* [2014] EWHC 2217 (TCC)/

6.106 The court or arbitrator is given a very broad discretion as to what amount of contribution to award. Section 2 of the Act provides:

2. Assessment of contribution. –

(1) Subject to subsection (3) below, in any proceedings for contribution under section 1 above the amount of the contribution recoverable from any person shall be such as may be found by the court to be just and equitable having regard to the extent of that person's responsibility for the damage in question.

6.107 What is "just and equitable" is a matter of causative potency and moral responsibility or, as is expressed in the contributory negligence cases, "blameworthiness".[101] This mix of considerations allows the court to take into account breaches of duty which were not causative of loss as well as those that were, albeit that causative breaches carry more weight.

6.108 In *Brian Warwicker Partnership plc v HOK International Ltd*,[102] architects and mechanical and electrical engineers were both engaged in a project to design and construct a shopping centre. The shopping centre had a central atrium which was intended to be kept at comfortable temperature. After the centre opened it was discovered that cold draughts entering though the automatic doors made it impossible to maintain these conditions. New doors had to be installed and the owners sued the engineers. The engineers compromised that claim and sought contribution from the architects. The court found that the architects had been in breach of duty to the owners in the design of the door systems and assessed contribution at 40%. Whilst primary responsibility lay with the engineers, a substantial part of the blame was attributable to the architects. In so doing the Judge took into account some aspects of the architects' conduct which were not causative of the damage. That finding was upheld on appeal, although the Court of Appeal warned against attributing too much weight to non-causative matters.

6.109 Whilst every case turns on its facts it should be noted that there is a "rule of thumb" for the division of responsibility between a specialist contractor and the construction professional which is responsible to the client for oversight and review of the contractor's work.

6.110 In *Carillion JM Ltd v Phi Group Ltd*,[103] the main contractor engaged a civil and structural engineer to advise in respect of engineering works and a specialist sub-contractor to carry out part of those works which extended to installing soil nails. The installation of the soil nails was defective because the contractor had misunderstood the ground conditions. The engineer was liable because it should have appreciated that the contractor's design was wrong. Akenhead J reviewed the authorities on contribution in cases of supervisors and persons supervised in the construction context, stating:

> The "poacher/gamekeeper" or the perpetrator/supervisor apportionment will often be in the 80–66.6% and 20–33% ranges respectively but where both contributors each have a

---

101 See *Madden v Quirk* [1989] 1 WLR 702 (CA).
102 [2005] EWCA Civ 962.
103 [2011] EWHC 1379 (TCC).

responsibility towards their mutual client to have regard to the same dangers and difficulties that does not seem to suggest that there is a poacher/gamekeeper scenario[104]

6.111 As between construction professionals, both of whom have a role in the production of some design or the occurrence of some event, the assessment of contribution will depend upon (1) which of their failings was more causative of the loss and (2) which of them owed the more onerous duties to the client (in respect of the facts which caused the loss).

6.112 In *J Sainsbury plc v Broadway Malyan*,[105] the architects designed inadequate fire-stopping for a supermarket. There was a fire and the architects were sued. The architects sought a contribution from the engineers. They were found not to be liable, but HHJ Humphrey Lloyd QC held that, had they been liable, their responsibility would only have been 12.5%. Their role was not to review or check the architect's design. It was to provide comments. On this basis the lion's share of responsibility lay with the architects.

---

104 On the facts, he found that the engineers were 40% responsible for the loss. Their role was more extensive than mere supervisor.
105 [1999] PNLR 268.

CHAPTER 7

Insurance

Introduction 139
Section A: summary 139
Section B: the contract of insurance 141
Section C: key aspects 144
   The proposal 144
   Notification 146
   Claims and aggregation 148
   Admissions 149
   Disputes 149
Section D: third-party rights 150

INTRODUCTION

7.1 Arguably the most important contract which any construction professional will enter into is its (annual) agreement for professional indemnity insurance. This form of insurance protects the construction professional against the consequences of claims against it for negligent breaches of duty.

7.2 For practising architects registered with the RIBA and practising surveyors registered with the RICS, carrying professional indemnity insurance to a minimum level is a condition of practice.[1] Many commercial clients will require proof of professional indemnity cover at a particular level as a condition of engagement, often further requiring that insurance be maintained at that level for a number of years after the project is completed.

7.3 When claims are made against construction professionals, it is the insurer of the construction professional which undertakes the defence of the claim and the level of insurance available is often a key consideration for any claimant pursuing a claim against a construction professional.

SECTION A: SUMMARY

7.4 All policies of professional indemnity insurance share core features.

7.5 Professional indemnity insurance covers the construction professional against claims made against it where the subject matter of the claim is some act or omission concerned with the construction professional's carrying out of its professional obligations.

7.6 The risk which the policy guards against is a claim arising from negligence on the part of the construction professional. For this reason, most policies will exclude claims

---

1 It is a requirement of the Architects Registration Board that practising architects carry a minimum level of PI cover of £250,000. The RIBA stipulate no minimum but require adequate professional indemnity cover. The RICS require a minimum level of cover of £1 million.

which arise from workmanship or the supply of goods or materials by the construction professional.

7.7 The scope of cover will attach to the construction professional's business. Thus if the construction professional trades through an LLP, the person described in the policy as "the Insured" will be the LLP. Most policies will go further. They will provide insurance cover to the construction professional in his or her personal capacity, and to the employees of the construction professional's business. Provided the claim made against these persons concerns negligence and arises out of the construction professional's business it will generally be covered.[2]

7.8 Each contract of insurance is annual. Every year the construction professional will obtain a policy of insurance for that year. Even if the construction professional remains with the same insurer and is insured on the same terms for a number of years, its insurance falls under sequential and separate policies.

7.9 The insurance is provided against a "claim". In contrast to household insurance or property insurance, professional indemnity insurance works on a "claims made basis". Thus a construction professional may carry insurance for a number of years during the life of a project which, particularly during the early years, will have very little practical bearing in respect of the risks of that project. What matters is the construction professional's insurance position when a claim arises. This can, of course, be during the life of the project, but it is much more likely in the years following completion.

7.10 This means that when a construction professional ceases to practice it will need to maintain "run off" cover for a minimum of six years.[3]

7.11 When an insurer agrees to provide a policy of insurance, the insurance it agrees to provide and the cost which it charges ("the premium") are all dependent upon the insurer's appreciation of risk. This means that in the negotiation of a policy of insurance the construction professional is required to provide the insurer with information relevant to the risk. This requirement can take the form of answers to specific questions and it can take the form of a general duty on the part of the construction professional to tell the insurer about facts or matters known to the construction professional which are likely to have a bearing on the insurer's appreciation of risk.

7.12 Failing to answer questions correctly or failing to provide material facts to the insurer may lead to the policy being "avoided" and/or the claim being declined.

7.13 It is frequently the case that facts or matters come to the attention of a construction professional which indicate that a claim may arrive in the future. Something may have gone wrong on the project and the construction professional may perceive a risk that the employer or some other person is likely to make a claim against the construction professional (even though it perhaps has not said so). Professional indemnity policies generally require this kind of "circumstance" to be "notified" to the insurer. If notification takes place and a claim subsequently materialises the claim is treated as falling within the policy current during the notification.

7.14 This mechanism is mainly intended to provide protection to the insured: if it did not exist, then any insurer negotiating a policy for subsequent years might decline to provide

---

2 On a strict analysis, professional indemnity policies are composite in nature: they are a bundle of contracts between different persons insured and the insurer. This *can* mean that some misrepresentation or breach of the terms by one insured does not rebound upon the other persons, see *Arab Bank plc v Zurich Insurance Co*. [1999] 1 Lloyd's Rep 262. In practice the insured committing the misrepresentation or breach is authorised to act for all the others and no defence arises.

3 Being the limitation period in contract. If the professional has signed contracts as a deed, the limitation period is 12 years (see above).

insurance at all or might provide it on prohibitive terms. It is also so that the insurer can take steps – should it so wish – to prepare for an anticipated claim, and insureds who fail to notify an insurer of a circumstance when they should do may find that a subsequent claim is declined because notification is a "condition precedent" to cover.

7.15 The financial value of cover is entirely a matter of what the insured wishes to pay. A construction professional in a modest way of business, working on smaller projects, may well opt for the minimum level of cover consistent with its regulatory requirements. The very largest practices will have insurance in the high tens of millions of pounds. Because professional indemnity insurance works on a "claims made" basis this level of cover will apply to each claim.

7.16 Thus a construction professional may have £5 million of cover, but if it is unlucky enough to face two claims of £4 million each falling into the same policy year they will both be met. The exception to this general statement is where the two claims share some unifying or common factor and the policy contains an "aggregation" clause which, because of that factor, treats the two claims as being one claim.

7.17 Insurance cover under professional indemnity policies will generally go beyond the financial consequences of a claim. Most importantly, such policies will generally indemnify the construction professional against the legal costs of investigating and defending a claim and against the legal costs of the person bringing the claim.

7.18 This fact, and the fact that the insurer indemnifies the construction professional against any damages awarded against it, have the effect that the insurer is accorded a substantial measure of control over the claim and over subsequent rights which arise. Thus the insurer will have control over whether and on what terms a claim is compromised and over whether and in what ways other persons are claimed against. If an insurer indemnifies a construction professional by making a payment to the claimant, the insurer will be entitled to recover the sum paid from a third party.

7.19 Sole practitioners and construction professionals in a modest way of business will often arrange their professional insurance directly with the insurer. However, larger concerns with more substantial insurance needs will generally work through an insurance broker. This form of agent is responsible for placing the construction professional's risk with an insurer in a way which is appropriate to the construction professional's needs. Where a broker in involved, all the construction professional's dealings with the insurer will be conducted though the broker.

7.20 The law of insurance is complex and a detailed examination of the way in which it applies to professional indemnity policies is beyond the scope of this book.[4] The discussion below is concerned with the areas of law and practice which are likely to be of particular significance to construction professionals.

SECTION B: THE CONTRACT OF INSURANCE

7.21 The contract of insurance is sometimes referred to as the policy of insurance, although this shorthand is not strictly accurate.[5] Like any contract it is made by offer and

---

4 The reader is referred to the leading textbook: *"Professional Indemnity Insurance":* Cannon & McGurk.
5 The contract is made up of a bundle of documents one of which contains the policy terms. The contract itself is constituted by the agreement of the insurer to provide the insured with insurance against liability in exchange for premium.

acceptance[6] and its terms are to be construed in the same way that any contract is construed. On renewal a new contract comes into being even if the terms are the same (including as to the premium). However, in terms of its shape and contents the written documents which constitute the contract of insurance are likely to unfamiliar to construction professionals.

7.22 *The Schedule*. A policy of professional indemnity insurance typically takes the form of standard policy terms to which are attached a schedule and various addenda. The Schedule sets out the bare essentials of the contract which are specific to this policy of insurance. The policy number will be given (each policy having its own unique number). The insured is identified (usually by reference to its trading name) and any subsidiaries if also covered. The period of insurance is identified. This will be an annual period, for example "from 1 October 2017 to 30 September 2018 both days inclusive". The Schedule will state the limit of indemnity, for example "GBP 5,000,000 any one claim including Defence Costs and Expenses". The excess will be stipulated. This is the contribution made by the construction professional to the cost of indemnifying any claim, for example "GBP 50,000". There may be further information. Usually the insurer is identified and there is a statement as to who any notification of a claim or circumstance is to be sent to (it is common for insurance to be provided by a number of insurers, one of which takes the lead). The "Proposal" (see below) and the premium may be identified.

7.23 *The Endorsements*. For larger contracts of insurance the insurer may have negotiated bespoke terms with the construction professional. This is particularly prevalent if the construction professional is represented by a broker. Some of these will serve to restrict the ambit of cover. For example, "The Policy excludes all claims which arise from the Insured's contracts with WindGen Ltd and/or any claim which arises from the provision of any service by the insured in relation to wind turbines". The may extend the ambit of cover, for example, "It is hereby noted and agreed that SYG designers are included within 'the Insured' for the purposes of this policy of insurance". They may clarify some uncertainty as to the scope of cover. For example:

> it is hereby noted and agreed that indemnity afforded by the Policy applies to liability attaching to the Insured arising out of a failure by the Insured to warn any client or clients of the inadequacy or deficiency (alleged or otherwise) in any design, specification or formula.

7.24 *The Standard Policy Wording*. The bulk of the documentation will consist of the standard terms and conditions which the insurer applies for this kind of professional indemnity insurance. This will generally follow broadly the same regime as follows.

7.25 *The Insuring Clause*. This is the key clause in the Policy. It describes the essence of the bargain between the insurer and the insured as to what claims will be covered. For example:

> The Insurer hereby agrees to indemnify the Insured up to the amount stated in the Schedule as the limit of indemnity for any sum which the Insured may become legally liable to pay arising from a claim or claims first made against the Insured and notified to the Insurer during the period of insurance and which arise out of the conduct by the Insured of its professional services.

---

6 Acceptance is judged objectively. Neither delivery of the policy nor payment of the premium is essential. However, what is essential is that acceptance be a clear intention on the part of the offeree to be bound and not a counter-offer. Because professional indemnity insurers operate according to standard procedures it is very unlikely that a dispute would arise as to whether a contract of insurance was entered into. It will almost always be clear from the documents.

7.26 Further clauses will then define "claim" and "professional services". The Insuring Clause may continue to extend indemnity to other types of loss, the most important being defence costs and the costs of the claimant, but it is common to find cover extended to such matters as "dishonesty or fraud" or "loss of documents" provided always these losses are sustained "arising out of the conduct by the Insured" of its professional services.

7.27 By the same token, sub-clauses of the Insuring Clause may restrict the scope of cover by making clear that the Insuring Clause has limitations in its scope (for example, liabilities which may arise by reason of a contractual agreement and which do not involve negligence). One of the most important restrictions is that the extent of indemnity will always be less than the excess.

7.28 *Exclusions*. All policies of insurance carry exclusions and these can be very important in professional indemnity policies. Thus it is relatively common to see an asbestos exclusion which is intended to exclude liability in respect of any claim "directly or indirectly arising out of or resulting from or in any way involving asbestos or any materials containing asbestos". One of the most important exclusions relates to claims arising from circumstances which should have been notified under a previous policy.

7.29 Typically the exclusion might state:

> The Insurer shall not be liable to indemnify the Insured … for any claim or arising from any circumstance known to the Insured prior to the inception of this policy or which the Insured should reasonably have known prior to the inception of this policy.

Policies will frequently exclude claims which fall to be indemnified under other policies of insurance (for example, any public liability policy obtained by the insured).[7]

7.30 *Conditions*. Separately from exclusions the Policy will contain a number of "Conditions". These may (depending on their true meaning) be conditions precedent to obtaining indemnity or conditions breach of which may permit the insurer to reduce its liability to the insured.[8] Where a condition is genuinely a condition precedent those words or words like them are generally employed.

7.31 One of the most important conditions precedent is the condition that any claim or circumstance be notified. This is generally coupled with the provision that a notified circumstance which results in a claim will fall within the policy under which the notification was made. These conditions are intended to prevent the situation where an insured properly informs prospective insurers of the risk that it might be sued and thereby becomes uninsurable against any future claim because the insurer either refused to insure at all, or will only insure on the basis that any future claim arising from that circumstance is excluded.[9]

7.32 Conditions which may not be conditions precedent may concern the insured's obligation to provide the insurer with cooperation in the event of a claim. Either here or In addition to conditions the policy may set out the regime under which the insurer will be

---

7 If that policy contains a similar exclusion, the Courts generally adopt the approach that the insured must be covered under either one or both.

8 Whether a condition is a condition precedent is very much a matter of the language used. A well written policy will either use the term "condition precedent" or will use words which plainly convey that meaning. In other cases it may be more difficult to ascertain the meaning. Generally the use of "must" or "shall" is a prerequisite for a condition precedent, but not of itself conclusive evidence. There is a special regime under the section 11 of the Insurance Act 2015 for ascertaining the insurer's rights in a situation in which conditions going to the avoidance of particular risks have not been complied with.

9 *Rothschild v Collyear* [1999] CLC 1697 (Comm).

entitled to conduct the defence of the claim against the insured and professional indemnity policies frequently provide both parties with the right to have any dispute as to whether the claim should be settled on specific terms referred to a neutral person, usually a Queens Counsel, for binding determination. More general disputes as to whether or not cover is afforded under the Policy will usually be referred to arbitration.

7.33 *Definitions.* Either at the end of the standard policy terms or at the beginning there will usually be a long list of definitions. These are important not just because of the specific meaning attached to particular words which play an important part in the mechanisms of the Policy (for example the meaning of "claim" or "circumstance"), but also because words which play an important part in the meaning of particular limitation of cover or exclusions will also be defined (for example "employee").

SECTION C: KEY ASPECTS

The proposal

7.34 When a construction professional applies for insurance to a new insurer it will be asked to prepare a statement of relevant facts about itself and its business which is generally referred to as "the Proposal". The Proposal will contain a number of statements about the business and specifically about the risk to the insurer.

7.35 One typical and very important statement is that the construction professional knows of no circumstances which might give rise to a claim against it other than those which it has disclosed in the statement. Where insurance is renewed by the insurer the construction professional is taken as having repeated those assertions as remaining accurate even if no "Proposal" has in fact been completed.

7.36 An insurance contract is a contract of utmost good faith.[10] That means, in particular, that there is a duty upon the insured to disclose to the insurer, prior to conclusion of the insurance agreement, all facts which are known by it that are material to the risk. As with any contract, if the insured makes a material misrepresentation the insurer may be able to reduce the scope of its obligations or decline them altogether. Since the Insurance Act 2015 came into force this means that the insured has an obligation to make a "fair presentation" of the risk. Section 3 states:

> (3) A fair presentation of the risk is one –
>   (a) which makes the disclosure required by subsection (4),
>   (b) which makes that disclosure in a manner which would be reasonably clear and accessible to a prudent insurer, and
>   (c) in which every material representation as to a matter of fact is substantially correct, and every material representation as to a matter of expectation or belief is made in good faith.
>
> (4) The disclosure required is as follows, except as provided in subsection (5) –
>   (a) disclosure of every material circumstance which the insured knows or ought to know, or

---

10 Marine Insurance Act 1906, section 17.

(b) failing that, disclosure which gives the insurer sufficient information to put a prudent insurer on notice that it needs to make further enquiries for the purpose of revealing those material circumstances.

7.37 Section 7(4) gives some examples of things that might be material: (a) special or unusual facts relating to the risk; (b) any particular concerns which led the insured to seek insurance cover for the risk and (c) anything which those concerned with the class of insurance and field of activity in question would generally understand as being something that should be dealt with in a fair presentation of risks of the type in question. But these examples are not exhaustive.[11]

7.38 Essentially, save where the insurer can be said to be on inquiry, the onus remains on the insured to divine what an insurer would regard as material, and thus in reality for many insureds the burden will remain upon their broker to give proper explanation and ask appropriate questions

7.39 Under the law prior to the 2015 Act, in the event that material information was not disclosed, or a material misrepresentation made, and the insurer could show that the failure to disclose that information, or the misrepresentation, induced it to enter into the contract (in the sense that it would not have made the same contract had it known the information or had the misrepresentation not been made),[12] it was entitled to avoid the contract. The premium was returnable and the insurer was relieved of all obligations to indemnify under that policy.

7.40 In response to a perception in the insurance market that material non-disclosure presented an unacceptable commercial risk to businesses such as construction professionals many insurers began to provide policies containing "innocent non-disclosure" clauses. Such clauses typically state that insurers will not avoid a policy (and/or decline liability) if the insured is able to prove to the insurers reasonable satisfaction that any non-disclosure or misrepresentation of material facts was not done deliberately.[13] Some clauses preserve the right of the insurer to reduce its liability in the event that, had it known the true facts, it would have offered insurance on different terms. It is now commonplace for professional indemnity policies to contain such clauses.[14]

7.41 A construction professional which, prior to inception of a policy, fails to provide material information or misrepresents material facts is thus at risk of losing some or all of the indemnity otherwise available under the policy. The extent of that risk will depend upon (1) the terms of the policy (and specifically whether there is an "innocent non-disclosure clause" and (2) the operation of the 2015 Act.

7.42 If there is no innocent non-disclosure clause, the 2015 Act regulates the rights of the insurer in this respect. The remedies available to an insurer for breach of the duty of fair presentation depends upon whether the breach was either (a) deliberate or reckless, or (b) neither deliberate nor reckless. Where the breach of duty was deliberate or reckless, the insurer may avoid the contract and refuse all claims. Deliberate non-disclosure or misrepresentation is rare.

---

11 Law Com No. 353, paragraph 6.13.
12 *Pan Atlantic Insurance Co Ltd v Pine Top Insurance Co Ltd* [1995] 1 AC 501 (HL).
13 See, for example, the wording offered by the RIBA Insurance Agency, condition 7 and Special Institution Condition 1 of the RICS Minimum Wording [2015].
14 See Rix LJ in *HIH Casualty and General Insurance v New Hampshire Co.* [2001] 2 Lloyd's Rep 161 at 205.

Particular in the world of construction professionals non-disclosure or misrepresentation tend to occur as a matter of accident or inadvertence (typically the person completing the proposal will have overlooked or will be unaware of a problem that has emerged on a project on which someone else in the business is engaged). Where the breach was neither deliberate nor reckless, the remedy will depend upon what the insurer proves it would have done had the presentation been fair.

7.43 Where it would not have been prepared to insure on any terms, the old remedy of avoidance will apply.[15] If the insurer would have been prepared to insure, but only on different terms (other than as to premium), the insurer may elect to treat the contract as if it had been entered into on those terms.[16]

7.44 In some cases this will be disastrous for the construction professional. If the problem which should have been disclosed was sufficiently serious the insurer may be able to show that it would have excluded any claim arising in respect of it from the scope of cover. Other less damaging consequences are that an increased excess would have applied, or that risk management conditions would have been incorporated into the policy. Because the Act has only been in force since August 2016, there are no reported cases illustrating how these provisions will work in practice.

7.45 It follows that particularly where one individual working in a construction professional business is given the responsibility for the insurance proposal it is particularly important that care is taken to check for:

- problems or potential problems which have arisen in respect of any project on which the business has been engaged which *might* result in a claim against the business at some time in the future;
- changes in the kind of work undertaken by the business which might result in the business working in an unusual environment or undertaking unusual obligations (for example, taking contractual responsibility for other designers practising in a different field);
- problems or unusual circumstances affecting the staff of the business and particularly its directors or senior partners (for example, has one of them been declared bankrupt).

7.46 This is not an exhaustive list. Where the construction professional works through a broker, the latter will be able to advise on which matters need to be disclosed and which do not. However, as a matter of general prudence, construction professionals should err on the side of caution where there is a doubt.

Notification

7.47 Notification of both claims and circumstances is a critical mechanism underpinning professional indemnity policies. Typically notification must be given promptly and often policies will set out requirements for specific types of information which are to be provided with the notification. As set out above, notification of circumstances is particularly

---

15 Schedule 1 paragraph 4. The premium will have to be returned.
16 Schedule 1 paragraph 5.

important so that the insured is not caught in the position of being uninsurable against future claims. A typical clause might state:

> The Insured shall as soon as reasonably practicable notify the Insurer of:
> (a) any claim made against the Insured;
> (b) the receipt of any notice from any person of an intention to make a claim against the Insured;
> (c) any potential loss or expense to be incurred by the Insured which is likely to give rise to a claim under this policy;
> (d) any circumstances known to the Insured which may reasonably give rise to a claim under this policy.
>
> Such notice having been given in accordance with the foregoing clause, any subsequent claim made after the expiry of the period of insurance specified in the Schedule arising from such circumstance or potential loss and expense or intention to make a claim shall be deemed to have been made during the currency of this policy.
>
> The due observance of the foregoing clause is a condition precedent to the Insurer's liability under this policy.

7.48 It would be unusual for a construction professional not to provide its insurer with prompt notice of a claim. A formal demand for compensation or an assertion that legal proceedings are likely to follow is generally sufficient to prompt the construction professional to consider its insurance position and notify its insurer.

7.49 Difficulties are generally encountered in the area of notification of "circumstances". Most policies leave the definition of "circumstances" very general. The important words are usually those governing the knowledge that the insured is required to have and the threshold of likelihood of that a claim will eventuate. Each case will depend upon the precise wording of this part of the policy,[17] but the following general observations may be helpful.

7.50 "Circumstances" are typically some kind of problem which could be explained in terms of being the result of a breach of duty on the part of a construction professional. Thus if the project is delayed because of a design clash involving the construction professional's design or because of a lack of coordination between some other person and the construction professional there is the possibility of a claim against the construction professional.

7.51 Whether the possibility is enough depends on the words used. Most policies provide that any circumstance which "may" give rise to a claim need to be notified. "May" means that there is a meaningful chance of a claim which can be less than 50%. Policies which state that the insured is notify circumstances which are likely to give rise to a claim, probably require a likelihood of more than 50%.[18] In each case the test is objective. The fact that the construction professional believed that there was no prospect of a claim (because it was convinced it had done nothing wrong) is irrelevant.

7.52 When a policy speaks of circumstances "known to the insured" this usually means circumstances known to the senior persons in the insured. A circumstance only known to a very junior person may not be sufficient. However, the fact that the management of a construction professional business did not in fact know about the circumstance because some senior person failed to tell them.

---

17 For an example of how the courts consider such problems, see *Kajima UK Engineering Ltd v Underwriter Insurance Co Ltd* [2008] EWHC 83 (TCC).
18 *Layher v Lowe* [2000] Lloyd's Rep IR 510 (CA).

7.53 Frequently problems occur as to what has been notified. An architect may be aware of a particular problem on a particular project and will give the insurer brief details. Does that amount to notification of circumstances relating to precisely the same problem, involving the same employee, on a different project (something that may not have been apparent at the time)? The courts have to balance divergent considerations. If a notification is construed too broadly it risks becomes meaningless, telling the insurer nothing of value. If a notification is construed too narrowly, the commercial purpose of notification may be subverted to pedantry.

7.54 Each of the decided cases turns upon the meaning of the words used,[19] but as a generality it is usually prudent to provide more information rather than less in any notification and in particular to indicate what is *not* known where the circumstances which may lead to a claim have been only partially investigated.

Claims and aggregation

7.55 Policies of professional indemnity insurance respond to "claims". Sometimes it is difficult to ascertain whether a package of complaints threatening legal action is a single claim or two or more claims combined.

7.56 In *Thorman v New Hampshire Insurance Company*[20] Sir John Donaldson MR gave a number of examples. If an architect has two separate contracts with two separate building owners and makes the same mistake in respect of both contracts there are two separate claims notwithstanding that (at least arguably) there is a single error. If the architect has a single contract in relation to two houses to be built on two different sites and in respect of one makes a mistake in relation to the windows, whilst in relation to the other it makes a mistake as to the foundations, there are again two separate claims. If, however, the architect has a single contract in relation to two different houses and makes it makes the same mistake in respect of both of them, there is a single claim. If the architect has a single contract in relation to a number of houses in a development in respect of which it has made a series of different mistakes, varying from house to house, there is none the less a single claim. However, in that situation it would matter whether the complaints were packaged together – in which case they might constitute a single claim – or were delivered sequentially over a number of years – in which case they might be treated as different claims.

7.57 Partly as a result of these potential uncertainties most professional indemnity policies contain an "aggregation" clause, the effect of which is to treat a number of claims which are sufficiently closely connected by a unifying feature as one claim. A typical clause might provide:

> All claims and losses resulting from:
>
> - one and the same act, error or omission; or
> - a series of acts errors or omissions arising out of the same originating cause or event,
>
> shall jointly constitute one claim under this policy, and only one excess will be applicable in respect of such claim

---

19 See, for example, *HLB Kidsons v Lloyds Underwriters* [2009] 1 Lloyd's Rep 8 (CA).
20 [1988] 1 Lloyd's Rep 7 (CA).

7.58 Aggregation clauses can work in favour of a construction professional, but they can also be detrimental. This can be demonstrated by examples. Suppose a civil and structural engineer is asked to design a number of different culverts for different clients involved in different road building projects. The engineer adopts the same methodology in each design and in each case the same design error is made. A number of different claims are made by different clients which fall into the same policy year. Individually none of them exceed £1 million. Collectively they total £8 million. If the engineer's limit of cover is £10 million and the claims are aggregated, the engineer will pay only one excess (rather than one excess for every claim). If the engineer's limit of cover is £5 million and the claims are aggregated the engineer may find itself uninsured for the value of the claims over that sum. Aggregation is thus an important consideration for any construction professional which provides services where there is a risk of the same error being replicated.

Admissions

7.59 It is part of the bargain between the construction professional and its insurer that the latter has conduct of the defence of claims. This necessitates restrictions upon the construction professional's freedom to deal with claims as it sees fit. For example, in respect of a relatively modest claim brought by an important client it may be in the construction professional's economic interests to admit responsibility early and make an offer of settlement. However, if that settlement is to the funded by the insurer both the admission and the offer must be made with the insurer's prior consent.

7.60 Most professional indemnity policies will contain a restriction along these lines and the consequence of the restriction being ignored is usually that the insurer is relieved from indemnifying the claim. A typical provision might state:

> For each and every claim the Insured and any person acting on behalf of the Insured must not admit responsibility, make an offer or promise, or offer payment or indemnity without the written consent of the Insurer.

7.61 Sometimes the construction professional and the insurer will disagree as to whether a claim should be settled. Most professional indemnity policies will provide a means of breaking that deadlock by referring the decision to a neutral third party for binding determination. Such a provision is generally referred to as a "QC clause" because the clause often stipulates that the person carrying out the determination will be a Queen's Counsel.

Disputes

7.62 In the vast majority of cases an insurer will appoint specialist solicitors to investigate the claim and to defend any proceedings brought against the construction professional. These solicitors will act for and will take their instructions from both the insurer and the construction professional, both being clients.

7.63 Sometimes the insurer will have a concern as to whether the claim falls to be indemnified at all. For example, there may be a difficult issue as to whether a condition precedent has been breached, or the issue as to whether the claim fails within the scope of the insuring clause may depend upon some factual matter which can only be determined at trial. Here the insurer will generally write to the construction professional stating that it is "reserving its rights". Both the insurer and the construction professional may then separately instruct their own solicitors to consider the issue of policy coverage, the solicitor instructed to conduct the defence being required to act for both of them. However, until such time as the insurer decides that the claim does not fall to be indemnified the insurer will continue to pay for that solicitor to investigate and defend the proceedings.

7.64 If the insurer declines to indemnify the construction professional, whether because it is entitled to avoid the policy pursuant to its terms and under the 2015 Act, or because the insurer takes the view that the claim falls outside of the scope of cover or is excluded (for example because of some failure to comply with a condition precedent), the construction professional will be left to defend the proceedings without assistance as to defence costs and without an offer of indemnity.

7.65 Plainly it is imperative that the construction professional obtains legal advice at this point if it has not already been obtained. The construction professional may be advised to commence proceedings against the insurer (which are usually in arbitration, which is confidential). Alternatively it may try to compromise the claim against it, or contest the claim against it, and then pursue the insurer. The mere fact that an insurer has declined cover may affect the dynamics of settlement (the usual consequence being that there is less merit in the claimant pursuing its claim as it may not recover its damages) but each case will turn on its facts.

SECTION D: THIRD-PARTY RIGHTS

7.66 A common feature of claims against construction professionals is that the construction professional becomes insolvent before the claim is made or shortly afterwards. In this instance the potential claimant faces a number of problems. Insolvency law in the United Kingdom can impose restrictions upon actions against insolvent persons, whether individual or corporate. More importantly, the fact of insolvency makes a claimant a potential creditor, with no greater rights than other unsecured creditors. The claimant may have no way of knowing whether there is a relevant professional indemnity policy and if so whether it would respond to the claim.

7.67 Since 1930, the harshness of this circumstance has been mitigated by statute so that acts of insolvency created a deemed assignment of the insured's rights to the claimant.[21] From 1 August 2016 the applicable regime is created by the Third Parties (Rights Against Insurers) Act 2010.

7.68 The Act confers the right to pursue an insurer directly if there has been an event of insolvency on the part of the construction professional. Insolvency means being subject to any one of a number of insolvency related orders or conditions.[22]

7.69 Investigations into an insolvent professional's insurance cover and requests for information about the identity of an insurer and an insurance policy can be made by a third

---

21 Third Parties (Rights Against Insurers) Act 1930.
22 Sections 4 to 7 inclusive. Thus in England and Wales a company is deemed to be insolvent for the purposes of the Act if it subject to a voluntary arrangement, an administration order, a winding up, the appointment of a receiver or the appointment of a liquidator.

party to an insurer, brokers, insolvency practitioners and former officers of an insolvent company prior to commencing court proceedings.

7.70 A request can be made for details of:

- the identity of the insurer;
- the terms of the policy;
- the limit of the cover;
- whether cover has been declined previously (and if so, details of any proceedings); and
- whether any aggregate of indemnity limit has been eroded.

7.71 A party can pursue a claim against an insurer without the need or requirement, in the first instance, to establish and obtain judgment that the insolvent construction professional is liable for the loss or damage.

7.72 This means that a single set of proceedings can be issued against the insurer to determine both the construction professional's liability and the potential liability of the insurer under an insurance policy.

7.73 The advantage of this is obviously one of cost savings as there is no longer a need for example, to take the step of restoring a company to the register of companies and/or to incur the cost of pursuing a claim against the construction professional in the first instance, and then subsequently commencing further proceedings against an insurer.

7.74 However, there is no requirement under the Act for an insurer to provide reasons for a refusal to indemnify. Reasons for a refusal to indemnify under an insurance policy may be obtained through receipt of a defence, requests for further information and/or disclosure but there is no guarantee that the information will be provided.

7.75 The Act does not put a third party in a better position vis a vis the insurer than the insured. This means that if the construction professional has lost its entitlement to cover under a policy of professional indemnity insurance there is nothing in the Act to undo that damage. It is a fact of commercial life that construction professionals, like other individuals and companies, sometimes neglect to follow good procedures or to pay bills when they are facing imminent insolvency. A construction professional which fails to notify a claim or a circumstance risks losing its entitlement to cover as does a construction professional which does not pay its premium leading to cancellation of the policy.[23] The Act does not suspend these obligations.[24] However, under section 9(2) the third party may do any act on behalf of the construction professional which is required to comply with a condition.

---

23 If the policy simply makes indemnity conditional upon payment of the premium that condition does not bite on the third party, although the insurer may deduct the premium from any payment.

24 It does operate to suspend the obligations of the construction professional to provide certain information during a period when the construction professional is insolvent.

CHAPTER 8

Experts, arbitrators and adjudicators

Introduction	153
Section A: experts in dispute proceedings	154
Duties and Part 35 of the Civil Procedure Rules	155
Retainer by instruction	158
Vulnerabilities	158
Section B: arbitrators	159
Appointment	160
Obligations	161
Vulnerabilities	162
Section C: adjudicators	163
Appointment	163
Obligations	164
Vulnerabilities	165
Section D: expert determination	165
Appointment	166
Obligations	167
Vulnerabilities	168

INTRODUCTION

8.1 Construction professionals are not only to be found working on projects, but also playing important roles in dispute resolution as expert witnesses, or deciding construction disputes as arbitrators, adjudicators or in expert determinations. In these roles the professional is in a quite different position from the traditional employer-consultant relationship, and each role brings different challenges (and risks).

8.2 Expert witnesses have an essential role in almost all professional negligence disputes. Only rarely can a claim for negligence against a construction professional properly be brought or defended without the input of expert witnesses from the relevant discipline.[1] Although a construction professional acting as an expert witness is instructed, and paid, by one of the parties to a dispute, their role is not simply to be a mouthpiece for that side (and experts who behave in such a way run the risk of being very publicly criticised by judges and having their evidence disregarded).

8.3 When appointed as arbitrators, adjudicators or in expert determinations, construction professionals are expected to decide disputes between parties. The use of people with industry-specific expertise to act as arbitrators has a long history, because the parties

---

1 In claims of limited financial value, the court may require the parties to appoint a single joint expert to save costs, rather than each party having its own expert. Even then, however, each party may have its own expert witnesses to advise on the claim behind the scenes.

hope and expect that such people will more easily understand the technical issues that might arise. However, the Housing Grants, Construction and Regeneration Act 1996 (see Chapter 9) has vastly expanded the opportunities for construction professionals to fulfil a decision-making role as adjudicators. Expert determination of disputes is much less common in the construction industry than litigation, arbitration or adjudication, but quite often used in particular kinds of dispute, such as rent reviews and measurement disputes.

SECTION A: EXPERTS IN DISPUTE PROCEEDINGS

8.4 As has been seen in Chapter 4, in any dispute that is likely to result in a claim that a construction professional has been negligent, an expert witness qualified in the appropriate discipline is likely to be needed at an early stage and almost certainly before such a claim can be pleaded. A construction professional accused of professional negligence (or, more likely, its insurer) will also want to retain an independent expert witness to assist with the preparation of a defence. This is necessary for the obvious reasons that, first, a defendant may find it difficult to deal dispassionately with the claim (and to see or accept that it has any merit) and, second, that in most cases a court or tribunal will lend more weight to the evidence of an independent expert witness than to the defendant who will, understandably, want to exculpate itself of blame. No matter what its level of expertise and experience, the defendant in most professional negligence claims must settle for being a witness of fact (that is, what it did and what happened) rather than being treated as an expert who can express an opinion as to whether the *Bolam* test has been complied with.[2]

8.5 This goes to the heart of the difference between "normal" and expert witnesses. Normal witnesses can give factual evidence, but are not generally allowed to express their opinion as to what conclusions should be drawn from those facts. That is because it is for the judge or tribunal to draw conclusions having first decided, on the balance of probabilities, what actually happened. By contrast, an expert witness gives evidence of its *opinion* as to what conclusions should be reached on the facts.

8.6 A simple example may assist. Suppose an engineer is accused of having negligently designed the piles for a warehouse. It is alleged by the claimant that this caused the floor of the warehouse to deflect and crack, necessitating repairs. In its defence, the engineer denies negligence and alleges that the damage was caused by the warehouse owner overloading the floor. The only evidence that the defendant engineer can personally give is as to the piling calculations that it performed and the inspections of the piling works that it undertook (if any). This is factual evidence. The engineer is not allowed to go on to say that, in its opinion, (a) a reasonable body of engineers would have done the same thing and (b) therefore, the engineer complied with its obligation of reasonable skill and care. This is opinion evidence, and the realm of the expert witness. The expert witness is there to give its opinion of professional practice and whether the issues with the floor were caused by the piling or are more likely to have been caused by something else (in this case, overloading by the owner).

8.7 Expert witnesses are required because judges and tribunals, no matter how experienced in construction matters, do not have the expertise needed to assess what constitutes

---

2 The position is not entirely hard-and-fast: see the judgment of Jackson J in *Multiplex Constructions (UK) Ltd v Cleveland Bridge UK Ltd (No. 6)* [2008] EWHC 2220 (TCC) at [657]–[676].

proper professional practice and to draw conclusions from technical information without assistance. As explained elsewhere,[3] the court or tribunal does not allow the expert to make decisions on its behalf but, having heard the expert evidence, forms a conclusion as to whether, on the balance of probabilities, the claim is made out.

8.8 The expert witness will be a key part of the parties' legal teams all the way up to and including trial, at which the expert will be required to give evidence and be cross-examined, and possibly beyond.[4] Typically, the expert will:

- Advise the legal team, who usually will not be qualified in the relevant construction discipline, as to the technical grounds for advancing or defending the claim. This information will be used for the Letter of Claim (or Response) and, in due course, for the pleadings.
- If the claim progresses, prepare an "expert report". This sets out the expert's opinion as to whether, on the facts, the claim shows that the defendant breached its duty of care and what the consequences were. Unsurprisingly, expert reports tend to reflect the case advanced by the instructing party. This is not (or, at least, should not be) mere partisanship: it should be obvious that, if a party cannot convince its own expert that its case is good, the claim is likely to be dropped or settle. Further, for reasons explained below, experts who simply act as advocates for their "side" run the risk of breaching their duty to the court and being severely and (and very publicly) criticised.
- The court will usually order that the experts for each party hold a "without prejudice" meeting (without lawyers present) to see if they can reach agreement on disputed matters within their area of expertise. An output of this meeting will be a list of agreed issues and issues where the experts do not agree, usually with comments from each expert as to why agreement has not been reached. This is encouraged by the courts because it is hoped that, by reducing the number of issues in dispute – or at least narrowing the issues that are disputed, the trial can be simplified with an associated reduction in time and costs.
- Possibly, prepare a supplemental report following the joint meeting, or if new information comes out.
- If necessary, give evidence at trial, including being cross-examined by the other parties' legal representatives.

Duties and Part 35 of the Civil Procedure Rules

8.9 The Civil Procedure Rules (the "CPR") are the rules of the civil (non-criminal) courts in England and Wales. They control all civil proceedings in the County Court and the High Court, where almost all professional negligence actions in England and Wales will be tried.[5]

---

3 See Chapter 4.
4 Expert witnesses are usually instructed by the parties' solicitors so that the instructions and all communications with the expert are subject to legal privilege (meaning that the other parties to the litigation are not entitled as of right to see these documents). A discussion of privilege is beyond the scope of this book.
5 Some professional negligence claims which result in breaches of health and safety lead to criminal prosecutions or, if a death has resulted, in an inquest. These proceedings are heard in different courts. Such proceedings

8.10 Part 35 of the CPR is concerned with expert evidence. CPR r.35.1 states:

> Expert evidence shall be restricted to that which is reasonably required to resolve the proceedings.

8.11 CPR r.35.4(1) says:

> No party may call an expert or put in evidence an expert report without the court's permission.

8.12 It follows that a party is not entitled to present expert evidence as of right: the court must give permission. If the court does give permission, it will also give directions as to the issues that the expert may give evidence on. This is to prevent parties (and experts) from producing expert evidence that is not relevant (or only tangentially relevant) to the claim and incurring unnecessary costs.

8.13 All this is mostly of interest to the lawyers, however. From the point of view of a construction professional who is asked to be an expert witness, the single most important rule in the CPR is 35.3:

(1) It is the duty of experts to help the court on matters within their expertise.
(2) This duty overrides any obligation to the person from whom the experts have received instructions or by whom they are paid.

8.14 This raises several fundamental points for an expert witness.

8.15 First, the primary role of the expert is to assist the court. By definition, an expert witness has knowledge about a specialist topic that the court will need help to understand: it is the experts' job to provide that help.

8.16 Second, the expert must only give evidence on matters within its expertise. This is a deceptively simple-seeming requirement, because it is easy to be drawn into commenting on matters of law or within the expertise of other disciplines. The expert witness should ensure that it has mastered all the relevant facts and has not proceeded on the basis of a misunderstanding of the facts or unjustified assumptions. For this reason, it is a requirement for the expert to include in his report the substance of his instructions and indicate where appropriate the documentary material upon which he has based his opinion. Legal proceedings concerning construction professionals often involve very substantial quantities of documentation and one of the tasks expected of an expert is that it will be able to master this documentation so as to enable it to provide a balanced and thorough opinion.

8.17 The expert must understand the nature of the function it is required to address. It is to provide evidence of technical matters and the standards adopted in its profession. It is not to provide an opinion as to what the expert personally would have done (although this is tempting). As HHJ Richard Seymour QC said in *Royal Brompton Hospital NHS Trust v Hammond (No. 7)*:[6]

> It is, in my judgment, essential for an expert witness in the trial of a professional negligence action to perform what is actually a very difficult task, at least unless one is experienced in doing

---

will almost always precede a civil claim for damages arising out of the events in question. However, both types of proceeding are outside the scope of this book.
6 (2001) 76 Con LR 148.

it, and that is to put on one side his own personal professional standards and to concentrate on the standards of the ordinarily competent member of his profession. There is a natural temptation to regard one's own standards as those which should be shared by all members of one's profession, but as those who are approached to act as expert witnesses are often approached just because they are especially prominent members of their profession or particularly experienced it is a temptation which must be resisted.

8.18 Third, experts must have sufficient expertise in the professional discipline to which their opinion is directed: it will be a rare case in which an architect can give relevant opinion evidence as to the standard of performance to be expected of a reasonably competent quantity surveyor; a generalist civil and structural engineer may not be qualified to provide relevant evidence in a case concerning the performance of a geotechnical engineer;[7] a non-practising engineer may find that its ability to provide evidence in respect of the standard of skill and care demanded from a practising engineer is put in doubt. In *Sansom v Metcalfe Hambleton & Co*,[8] a building surveyor failed to draw his client's attention to a crack in a wall which, it was alleged, indicated structural defects. The judge at first instance preferred the claimant's evidence that this was negligent, even though that evidence was provided by structural engineer and the defendant's evidence came from a building surveyor. The Court of Appeal held that this finding could not stand. Butler-Sloss LJ said:

> A court should be slow to find a professionally qualified man guilty of a breach of his duty of skill and care towards a client (or third party) without evidence from those within the same profession as to the standard expected on the facts of the case and the failure of the professionally qualified man to measure up to that standard.

8.19 Finally, and most importantly, CPR 35.3(2) makes it clear that the expert's primary duty is to the court, not to the party who has instructed and paid the expert. The expert, although paid by one of the parties, is really supposed to be working for the court. There are countless examples of experts falling foul of this rule and accidentally appearing to act as advocates for their instructing client instead of as an independent expert. In litigation tensions can run high and positions become entrenched: it is very easy for an expert to feel that concessions – even properly made – somehow undermine the expert's credibility. In fact, the opposite is true: an expert who in court is prepared to accept that its opinion would change given different facts, or that there may be an alternative explanation for the events in question, is much more likely to be trusted by the court than one who doggedly sticks to his "case" even in the face of the evidence.[9]

---

7 For an example of a case where these issues were discussed in an engineering context see *Cooperative Group Ltd v John Allen Associates Ltd* [2010] EWHC 2300.

8 [1998] PNLR 542 (CA).

9 In *National Justice Compania Naviera SA v Prudential Assurance Co Ltd (The Ikarian Reefer)* [1993] 2 Lloyd's Rep 68 at p. 81, Cresswell J set out the duties of expert witnesses more comprehensively than in the CPR. This is regarded as the leading exposition of experts' duties and had repeatedly been cited with approval by other courts. Cases where expert witnesses have fallen below the standards expected are numerous. Judges are, if anything, becoming more willing publicly to criticise such failures. For three recent examples in the TCC, see: *Imperial Chemical Industries plc v Merit Merrell Technology Ltd* [2017] EWHC 1763 at [60]–[86]; *Riva Properties Ltd and ors v Foster and Partners* [2017] EWHC 2574 at [51]–[59]; and *Bank of Ireland v Watts Group (UK) plc* [2017] EWHC 1667 (TCC) at [58]–[70].

8.20 Further guidance for expert witnesses is provided by the Civil Justice Council's *"Guidance for the Instruction of Experts in Civil Cases"*, which is available on the CJC's website.[10]

Retainer by instruction

8.21 Expert witnesses are not usually appointed using standard contract terms but by means of a letter of instruction from the client's solicitor. Such a letter is most unlikely to set out the expert's obligation of reasonable skill and care, but the obligation will be implied in the usual way.[11] An appointment as an expert witness is a personal one, that is to say that an individual is appointed to act, not his firm or company (although the fees may be not be payable directly to the individual).

8.22 In addition to the duties of an expert witness set out above, the expert will need to confirm that he does not have a conflict of interest (that is, he has not been consulted by another party to the dispute, and does not have any interest in the subject matter of the dispute), that he has the time and expertise to discharge his obligations properly. Most reputable construction professionals acting as expert witnesses will hold professional indemnity insurance, but this is not usually a requirement of the instruction, and not all policies will cover this as standard. If in doubt, it is best to be cautious and check with the broker or insurer before agreeing to act.

Vulnerabilities

8.23 Historically, an expert witness was immune from having proceedings brought against them arising out of their performance giving evidence in court. The justification for that was that experts should not feel constrained by their client's best interests when providing their evidence.

8.24 This "immunity from suit" was abolished by the Supreme Court in *Jones v Kaney*.[12] Lord Phillips PSC said that it was not immunity from suit that compelled the expert to give frank evidence to the court, but integrity. That, he felt, had nothing to do with the possibility of the expert being sued by its client:

> An expert's initial advice is likely to be for the benefit of his client alone. It is on the basis of that advice that the client is likely to decide whether to proceed with his claim, or the terms on which to settle it. The question then arises of the expert's attitude if he subsequently forms the view, or is persuaded by the witness on the other side, that his initial advice was over-optimistic, or that there is some weakness in his client's case which he had not appreciated. His duty to the court is frankly to concede his change of view. The witness of integrity will do so. I can readily appreciate the possibility that some experts may not have that integrity. They will be reluctant to admit to the weakness in their client's case. They may be reluctant because of loyalty to the client and his team, or because of a disinclination to admit to having erred in the initial opinion. I question, however, whether their reluctance will be because of a fear of being sued-at least a fear of being sued for the opinion given to the court. An expert will be well aware of his duty to the court and that if he frankly accepts that he has changed his view it will be apparent that he is performing that duty. I do not see why he should be concerned that this will result in his

---

10 www.judiciary.gov.uk/related-offices-and-bodies/advisory-bodies/cjc/cjc-publications/guidance-for-the-instruction-of-experts-in-Civil-claims/
11 A term will be implied pursuant to the Supply of Goods and Services Act 1982, see Chapter 3.
12 [2011] UKSC 13.

being sued for breach of duty. It is paradoxical to postulate that in order to persuade an expert to perform the duty that he has undertaken to his client it is necessary to give him immunity from liability for breach of that duty.

8.25 This change to the law does not, as yet, appear to have resulted in a flood of claims against construction professionals for negligently giving evidence in court. Lord Phillips suggested that one of the beneficial consequences of removing immunity from experts might be that they would be more conservative in their initial advice to their clients, having regard to the possibility of being sued if it transpired in court that the advice was wildly optimistic.

8.26 In any event, expert witnesses have always had liability to their clients for their initial advice. At the start of a case, the client will be relying on the expert's advice when it decides whether to pursue, defend or settle a claim. At that stage of proceedings, the expert is in a similar role to any professional who is being asked to give a second opinion on another professional's work. A failure by the expert to use reasonable skill and care in giving that advice may give rise to a claim for the wasted expenditure in pursuing litigation that might otherwise have settled (or settled for a lesser sum). Such claims are not common, and not without their difficulties (particularly in causation).

8.27 Probably the greatest risk to a construction professional who agrees to act as an expert witness is reputational. Court proceedings are almost always held in public and, in the High Court and above, the judgments are given in public and widely published. It is a very public forum in which to be exposed. Judges generally accept that giving evidence and being cross-examined is a stressful and difficult experience and they do not criticise experts for this. However, experts who do not come up to the standard expected of them – as described above – run the risk of severe and very public criticism from the judge.[13]

SECTION B: ARBITRATORS

8.28 Whenever parties to an arbitration agreement refer a dispute under that agreement to arbitration, one or more arbitrators will have to be appointed as the tribunal. The precise number of arbitrators depends on the terms of the arbitration agreement, but is usually one (known as a "sole" arbitrator) or three. Both sole and multiple arbitrators are referred to as the "tribunal". Typically, where a tribunal consists of three arbitrators the parties will choose one arbitrator each and those arbitrators will then select the third arbitrator, who acts as the chairman of the tribunal.

8.29 The role of the arbitrator is to decide the dispute referred by the parties. It is analogous, but not identical, to the role of a judge in litigation. The main differences between an arbitrator and a judge are (1) the arbitrator's authority comes from the parties, not the state; (2) the arbitrator sits in private; and (3) arbitrators lack many of the coercive powers that judges have to compel compliance with directions and awards.[14]

---

13 In addition to the cases referred to above, see the case of *Van Oord UK Ltd and anor v Allseas UK Ltd* [2015] EWHC 3074 (TCC) at [80]–[94]. It is a salutary warning to experts and would-be experts as to the importance of properly preparing, mastering the material and understanding what the role requires.

14 In arbitrations to which the Arbitration Act 1996 applies, the court has powers that can be exercised in support of orders that the tribunal may make: see sections 42–44. This, however, requires a party or even the tribunal itself to make an application to the court.

8.30 The use of construction professionals as arbitrators is well established. *Russell on Arbitration* (24th Edition) states that "it is a feature of English arbitration practice that non-lawyers may become arbitrators in specialist fields such as rent reviews, engineering, shipping and construction".[15] However, it is increasingly common for non-lawyers to act as arbitrators in international arbitrations in these specialist fields too. Especially where the subject matter of a dispute is highly technical, there is a strong perception that it can be beneficial to have one or more experienced, technically qualified arbitrators as part of the tribunal. Many such arbitrators also hold legal qualifications although they are not qualified to practise as lawyers. A number of construction professionals practise almost solely as arbitrators, and the Chartered Institute of Arbitrators (among others) runs courses and sets examinations for people who wish to become accredited.

Appointment

8.31 The precise details for the appointment of an arbitrator depends upon the terms of the arbitration agreement between the parties. As the legitimacy of the arbitration process is derived from the consent of the parties, any provisions in the arbitration agreement must be followed precisely. A failure to do so risks invalidating the appointment, robbing the arbitrator of jurisdiction and rendering the arbitration a nullity. It will be rare for a construction professional to be involved in this process, save as the person being appointed as the arbitrator.

8.32 Typically, in an arbitration to which the Arbitration Act 1996 applies[16] and where the tribunal is to be a sole arbitrator, a party starts an arbitration by sending the other party a "notice of arbitration". This will set out the basis of the dispute, the key facts upon which it relies, and a description of what remedy is sought. The notice will either incorporate or be accompanied by a proposal to agree to a specific person as the arbitrator (usually a choice of possible arbitrators will be offered, or the arbitration agreement will name specific candidates).

8.33 If the parties are able to agree to the identity of an arbitrator,[17] the next step will be to "refer" the dispute to that person, along with a request that they act as arbitrator. Assuming that person is prepared to act, they will send their terms and conditions for doing so to the parties for signature. This is often a brief document, setting out no more than the arbitrator's hourly and daily rates for working on the arbitration (often including the right to payment on account of fees), terms for the arbitrator's expenses and a right to be paid whether the arbitration runs to its completion or not. For reasons explained below, the terms are usually silent as to the arbitrator's duty of care to the parties. Once signed by the parties, this forms a contract between them and the arbitrator, who will then give directions for the conduct of the arbitration itself.

---

15 At 4–011.

16 Subject to a few limitations, in England and Wales the parties are free to agree to use whatever arbitration rules they choose, so the details will depend on the content of the chosen rules. Rules are promulgated by various specialist bodies and arbitration "courts" such as the London Court of International Arbitration ("LCIA") and the International Chamber of Commerce ("ICC"). In construction arbitrations in the UK, the Construction Industry Model Arbitration Rules ("CIMAR"), published by the Joint Contracts Tribunal, are commonly used.

17 It is not uncommon for parties to be unable to agree anything between them. In that case, ss.17 and 18 of the Arbitration Act provide for the procedure in default of agreement. Bodies such as the LCIA and ICC also have procedures for appointing arbitrators in default of agreement between the parties.

Obligations

8.34 An arbitrator has many obligations to the parties and a complete exposition of all of them is beyond the scope of this book. Only the most significant are dealt with below.[18]

8.35 Ultimately, the arbitrator's task is to decide the dispute between the parties. However, the arbitrator has other obligations to the parties that must be discharged in reaching that decision. The most important are set out in s.33 of the Arbitration Act 1996:[19]

(1) The tribunal shall –
    (a) act fairly and impartially as between the parties, giving each party a reasonable opportunity of putting his case and dealing with that of his opponent, and
    (b) adopt procedures suitable to the circumstances of the particular case, avoiding unnecessary delay or expense, so as to provide a fair means for the resolution of the matters falling to be determined.
(2) The tribunal shall comply with that general duty in conducting the arbitral proceedings, in its decisions on matters of procedure and evidence and in the exercise of all other powers conferred on it.

8.36 Acting "fairly and impartially" between the parties is not confined only to allowing each party to put its case and respond to the case against it. It includes making directions that are fair to both sides and an obligation of candour in dealing with the parties, particularly with respect to possible conflicts of interest and communication with the parties.

8.37 Conflicts of interest can arise more easily for a "professional" arbitrator than might be expected, not least because it is possible in the relatively small world of construction for an arbitrator to have heard multiple cases involving a particular party. The Chartered Institute of Arbitrators' Code of Professional and Ethical Conduct for Members attempts to address this issue at rule 3, which states:

> Both before and throughout the dispute resolution process, a member shall disclose all interests, relationships and matters likely to affect the member's independence or impartiality or which might reasonably be perceived as likely to do so.

8.38 Anyone considering acting as an arbitrator would be well advised to take this guidance seriously. Parties are increasingly aware of the possibility of conflicts arising and willing to apply to the courts for orders removing arbitrators in appropriate circumstances.[20] A good rule of thumb if considering whether something might be a possible conflict of

---

18 For what must be regarded as an exhaustive list and explanation of the duties of an arbitral tribunal, see *Russell on Arbitration* (24th Edition), at 4–109 *et seq*.

19 Similar obligations are imposed on tribunals by other arbitration rules. See, for example, rules 14.4 and 5.3 of the LCIA Rules 2014. Even if the rules are silent on this, in an arbitration in England, Wales or Northern Ireland the obligation would still be imposed by virtue of ss.2 and 33 of the Arbitration Act 1996.

20 The court has power to do this under s. 24 of the Arbitration Act 1996. This power is rarely used. A recent example is *Cofely Ltd v Anthony Bingham and Knowles Ltd* [2016] EWHC 240 (Comm), where in the preceding three years the arbitrator had received 18% of his appointments and 25% of his fee income from cases brought by Knowles. Knowles had previously been found to be fraudulently manipulating adjudication appointments to obtain particular adjudicators (see *Eurocom Ltd v Siemens plc* [2014] EWHC 3170 (TCC)). When Cofely inquired about the relationship between the arbitrator and Knowles, the arbitrator refused to disclose the requested

interest (or might reasonably be perceived as being one by either of the parties) is to err on the side of disclosing the matter.

8.39 Arbitrators should not assume that the parties will try to make it easy for them to discharge the duty of fairness and impartiality. Although arbitration is "consensual" in the sense that the parties have agreed to submit their dispute to the process, it is every bit as adversarial as litigation and the parties (or their lawyers) will often seek to make life as difficult as possible for each other in order to gain a tactical advantage. The arbitrator has to be robust in dealing with this behaviour while still being fair to both sides.

8.40 Making a decision in the dispute (called an "award") is not as simple as picking which side has won and ordering payment (or whatever other relief is sought) accordingly. The arbitrator is required to give reasons for the decision that is reached,[21] must deal with all the issues put by the parties and must not exceed the scope of the dispute referred. This can be a difficult and onerous task. An award that does not comply with these requirements is open to challenge in the courts, as is an award that is based on an error of law (although challenges on points of law are relatively rare).

8.41 In reaching a decision, the construction professional acting as an arbitrator must resist the temptation to rely on reasoning derived from its own experience and expertise without first giving the parties an opportunity to make submissions on the point in question. A failure to do this is also grounds for an award to be set aside.

Vulnerabilities

8.42 Under the Arbitration Act 1996, arbitrators enjoy complete immunity for their actions as arbitrator, save in cases of bad faith.[22] "Bad faith" is not defined in the Arbitration Act, nor is it well-defined in common law (where it is often equated with fraud). Certainly, it would seem that some element of dishonesty or deliberate subversion is likely to be required. Thus, in *Webster v Lord Chancellor*[23] Sir Brian Leveson cited with approval a passage from *Cannock Chase District Council v Kelly*:[24]

> I would stress – for it seems to me that an unfortunate tendency has developed of looseness of language in this respect – that bad faith, or, as it is sometimes put, "lack of good faith," means dishonesty: not necessarily for a financial motive, but still dishonesty. It always involves a grave charge. It must not be treated as a synonym for an honest, though mistaken, taking into consideration of a factor which is in law irrelevant.

8.43 There are no recorded cases of arbitrators being sued for professional negligence, and it is unlikely that mere negligence could amount to bad faith. An arbitrator who is shown to have acted in bad faith (no easy task) will almost certainly be removed by the court under s.24 of the Act and is unlikely to be able to recover its fees (in addition to being sued by one or both of the parties).

---

information. Hamblen J found that apparent bias had been made out and would have ordered the arbitrator's removal if he did not resign first.

21 Unless the parties agree otherwise, which will almost never happen: Arbitration Act 1996, s 52.
22 Arbitration Act 1996, s 29.
23 [2015] EWCA Civ 742 at [30]. A number of other authorities were also referred to on this point.
24 [1978] 1 WLR 1 (CA).

8.44 As with expert witnesses, and despite the fact that the arbitration process is confidential to the parties, the biggest risk for an arbitrator is reputational. Having one's name associated with litigation about apparent bias, or other failures in the conduct of an arbitration is not good publicity. Where the arbitrator is also a member of a professional body that nominates arbitrators (such as the CIArb or the RICS), there is also the possibility of being subject to disciplinary measures after the court proceedings have finished. Even where difficulties are resolved without publicity, the construction industry is a small one where it is easy for a reputation to be tarnished.

SECTION C: ADJUDICATORS

8.45 Unlike arbitrators, who have a long history in English law going back to at least the thirteenth century, adjudicators are a modern phenomenon, having been created by Parliament in 1996. An adjudicator is appointed by the parties to decide an adjudication, whether brought under the statutory scheme or under the terms of a contract.[25] They are almost uniquely active in construction disputes (although there is no reason why parties cannot incorporate adjudication into contracts about other matters).

8.46 Construction professionals have established themselves as probably the main body from which adjudicators are drawn. This is in no small part to the fact that the construction professional bodies such as the RICS and the RIBA have become the most popular nominating bodies for appointments (partly by being included in many standard form contracts). It is also likely that, as many adjudications are relatively self-contained and about payment disputes, construction professionals are seen as being at least as well-suited to deciding such disputes as lawyers.

8.47 As explained in Chapter 9, adjudication applies to almost all contracts for construction operations, a term that is widely defined and encompasses most appointments that construction professionals will enter into (save in the case of certain residential works). This includes contracts that are not recorded or evidenced in writing.

Appointment

8.48 In contrast to arbitration, it is relatively uncommon for adjudicators to be named in the contract that gives rise to the dispute and parties do not often agree the identity of the adjudicator. The usual course is for the party that wishes to commence an adjudication to ask a recognised adjudicator nominating body (known as an "ANB") to nominate a person on the ANB's list to act as the adjudicator. There are many ANBs including the RICS, the RIBA, TeCSA (the Technology and Construction Solicitors' Association), TECBAR (the Technology and Construction Bar Association) and the CIArb. Usually the ANB that must be used is named in the contract.

8.49 Different ANBs have different procedures but, generally, having received a request for a nomination, the ANB will select a potential adjudicator and refer the dispute to that person, who must then confirm whether they are willing to act. This person must confirm that they do not have any conflict of interest and will then write to the parties to set out the terms on which they are willing to act (including terms as to their fees). Once the

---

25 See Chapter 9.

parties have confirmed their acceptance of these terms, the adjudication proceeds. It is not uncommon for the defendant (called the "responding party") to remain silent at the point of appointment, or at least to reserve its position. This is generally a tactical device while the responding party decides whether it has to participate in the adjudication at all.

8.50 The appointment forms a contract between the parties and the adjudicator. The importance of the appointment was highlighted in the case of *PC Harrington Contractors Ltd v Systech International Ltd*,[26] where an adjudicator's decision was ruled to have been a nullity (the reasons for that are not important for instant purposes). The Court of Appeal held that in those circumstances there had been a total failure of consideration on the part of the adjudicator (that is, the parties had received nothing of value from the adjudicator) and could therefore not recover any payment for his work.

8.51 This is obviously an unattractive position for an adjudicator to be in. The adjudicator has to reach a decision in a short space of time, and the parties will rarely make the adjudicator's life easier by clearly setting out the boundaries of the dispute and the procedure that they wish to follow. Indeed, for most responding parties, it will be tactically attractive to put the adjudicator in doubt as to the extent of his jurisdiction.

8.52 The answer to this problem, which was identified by Davis LJ in *PC Harrington*, is for the construction professional to ensure that its terms of appointment provide that it will be paid for its time spent on the adjudication regardless as to whether the decision is enforceable or not.

Obligations

8.53 As with arbitrators, it is beyond the scope of this book to set out all of the obligations that adjudicators have to the parties.[27] An adjudicator is in a comparable position to an arbitrator, with a similar statutory responsibility to act impartially as between the parties.[28] Accordingly, all that has been said of arbitrators' duties at paragraphs 8.34 to 8.41 above can equally be applied to adjudicators.

8.54 The courts are alive, however, to the special pressures that adjudicators are under in having to deliver a decision in a very short space of time. To prevent parties from undermining the purpose of adjudication, the courts deal robustly with challenges to the validity of adjudicators' decisions and allegations that there has been a breach of natural justice (fairness and impartiality) or jurisdictional errors (essentially, questions as to whether the adjudicator had the power to reach the decision that it did). More leeway is therefore given to adjudicators than to arbitrators and, even if the adjudicator reaches an incorrect decision, the court will generally enforce it:

> unless it is plain that the question which [the adjudicator] has decided was not the question referred to him or the manner in which he has gone about his task is obviously unfair... The need to have the right answer has been subordinated to the need to have an answer quickly.[29]

---

26 [2012] EWCA Civ 1371.
27 There is little material on how adjudicators should conduct themselves. A good introduction can be found in *Coulson on Construction Adjudication* (Third Edition) at Chapters 18 to 20.
28 Housing Grants, Construction and Regeneration Act 1996, s 108(2)(e).
29 Per Chadwick LJ in *Carillion Construction Ltd v Devonport Royal Dockyard Ltd* [2005] EWCA Civ 1358 at [85]–[86].

8.55 This does not amount to a complete freedom for adjudicators. The warnings given about conflicts of interest apply equally to adjudicators, where these questions have arisen on a number of occasions.[30] Adjudicators must be conspicuously impartial: unilateral communication with one of the parties (which can be tempting when time is so short) should be avoided also.[31]

8.56 It is particularly important for adjudicators to provide reasoned decisions that take into account the parties' arguments and which do not rely on the adjudicator's own theories (based upon their specialist knowledge) if those theories have not first been put to the parties for their submissions. This is because, once the 28 day time limit has expired, there is no jurisdiction for the adjudicator to go back and correct (save for minor "slips"), or add to the decision. A defective decision will almost always be unenforceable.

Vulnerabilities

8.57 Again, as with arbitrators, adjudicators benefit from a statutory immunity for acts and omissions carried out in their role, save to the extent that such act or omission was in bad faith.[32] There is no recorded case where an adjudicator has been sued for professional negligence. The usual course for parties that consider that an adjudicator has not performed its obligations is for them to refuse to pay its fees.[33]

8.58 The principal risk to construction professionals (and others) acting as adjudicators is also reputational. Much of the business of the Technology and Construction Court of the High Court concerns enforcement of adjudicators' decisions. Decisions that are not enforced generate considerable coverage in the legal and construction press (because it is comparatively rare for the TCC to refuse enforcement). In recent years, the court has tended to name the adjudicator involved and, although the court rarely criticises the adjudicator directly, this cannot be described as "good publicity".

SECTION D: EXPERT DETERMINATION

8.59 Expert determination is, as the name suggests, a dispute resolution process whereby the parties contractually agree to refer their dispute to an expert third party for a binding decision. It is not a common form of dispute resolution, and is most often used for valuing shares in business transactions. In the context of construction disputes, it is most commonly used to settle disputes about measurement of usable space (which is obviously apt

---

30 See, for example, *Beumer Group UK Ltd v Vinci Construction UK Ltd* [2017] EWHC 2283 (TCC). In that case the adjudicator had been adjudicating two disputes involving Beumer at the same time, which related to the same project. Beumer was running factually different cases in the two adjudications. Fraser J held that this should have been disclosed to Vinci.

31 See *Paice and Springall v MJ Harding Contractors* [2015] EWHC 661 (TCC), where one of the parties had undisclosed discussions with the adjudicator's wife, who was also his office manager. As Fraser J observed in *Beumer*, unilateral telephone conversations should be regarded as "strongly discouraged (if not verging on prohibited)". It seems to be a recurring problem: see *Amec Capital Projects Ltd v Whitefriars City Estates Ltd* [2005] EWCA Civ 1358 and *Discain Project Services Ltd v Opecprime Developments Ltd (No. 1)* [2000] BLR 402.

32 Housing Grants, Construction and Regeneration Act 1996, s 108(4).

33 It should be noted that an adjudicator cannot make giving his decision conditional on being paid first. A failure to issue the decision within the statutory time results makes the decision unenforceable: *Mott MacDonald Ltd v London & Regional Properties Ltd* [2007] EWHC 1055 (TCC).

for determination by a suitably qualified surveyor), but there is no reason in principle why it cannot be deployed for other disputes.

8.60 Outwardly, a person making an expert determination appears to be fulfilling a similar role to an arbitrator or adjudicator. They may receive submissions and evidence from the parties, hold hearings, ask questions and then proceed to make a decision. The role is, however, fundamentally different: an expert is not an arbitrator (or adjudicator) and is not subject to the same statutory and judicial oversight. As will be seen, the decision made in an expert determination cannot be impeached except in very limited circumstances. That is commercially attractive because it provides parties with certainty: it limits the scope for appeals against or collateral attacks on the decision.

Appointment

8.61 Few contracts provide that all disputes that arise are to be settled by expert determination. The usual position is that only specific disputes are to be referred to an expert.[34] In addition to defining the which disputes can be referred, the expert determination clause will usually also specify:

- that the expert will decide the dispute as "an expert and not an arbitrator"; and
- the qualifications that the expert must hold. For example, for measurement disputes, the expert may be required to be a Fellow of the RICS with not fewer than 10 years' experience.

8.62 A well-drafted expert determination clause will also provide for a third party (usually the head of the professional body to which the expert is required to belong) to select someone to act in the event that the parties cannot agree.

8.63 A construction professional who is asked to act as an expert for the purposes of an expert determination will typically be appointed on the basis of a joint letter of appointment sent by the parties to the dispute. The letter should set out the usual terms as to remuneration and, most importantly, be clear as to:

- what the expert's instructions are as to the dispute to be decided; and
- the capacity in which the expert is being appointed.

8.64 The appointment will usually be silent as to the expert's obligation to use reasonable skill and care in carrying out his tasks, which will be implied in the usual way. As will be seen below, a construction professional carrying out an expert determination can, potentially, be liable for doing so negligently – there is no immunity as in the case of arbitrators or adjudicators. This means that a professional agreeing to act may wish to incorporate (or

---

34 Although most contracts that provide for expert determination expressly state that the decision will be "final and binding" on the parties, the presumption is that this is the case whether those words are used or not. However, non-binding determinations are possible, and as expert determination is a matter of contract between the parties, each case will depend on the precise wording of the contract in question. See *Homepace Ltd v Sita South East Ltd* [2008] EWCA Civ 1 at [18].

at least try to incorporate) a term into the appointment that excludes liability for anything done or not done while carrying out the expert determination, save where the act or omission is in bad faith. The fact that the expert has a potential liability in negligence to the parties means that the professional should ensure that its professional indemnity insurance provides cover for this work.

Obligations

8.65 The expert's principal obligation is only to make a determination that is in accordance with the instructions it has been given by the parties. If the instructions require the expert to give reasons for the decision[35] then it must do so, but those reasons do not necessarily need to be as detailed as would be the case in an arbitration or adjudication. Indeed, the reasons can be very brief.[36]

8.66 The importance of compliance with the expert's instructions cannot be overstated because a failure to do so will render the decision unenforceable. Thus, the decision of an expert who valued a piece of machinery himself, when he had been instructed to employ another expert to value it, was held to be void.[37] Similarly, if a surveyor is instructed to make an expert determination of the usable area of a property using SMM6, but proceeds to use SMM7 instead, the decision will be a nullity. If the instructions require the expert to take certain matters into account, then it must do so, even if it would not otherwise have done and even if it considers those matters irrelevant.

8.67 The position was helpfully summarised by Knox J in *Nikko Hotels (UK) Ltd v MEPC plc*:[38]

> if parties agree to refer to the final and conclusive judgment of an expert, an issue which either consists of a question of construction or necessarily involves the solution of a question of construction, the expert's decision will be final and conclusive and therefore not open to review or treatment by the courts as a nullity on the ground that the expert's decision on construction was erroneous in law, unless it can be shown that the expert has not performed the task assigned to him. If he has answered the right question in the wrong way, his decision will be binding. If he has answered the wrong question, his decision will be a nullity.

8.68 The reason for this is that the parties have agreed in their contract as to what they are prepared to submit to the expert and on what basis. The court will hold them to that contract. As Lord Denning MR said in *Campbell v Edwards*:[39]

> It is simply the law of contract. If two persons agree that the price of property should be fixed by a valuer on whom they agree, and he gives that valuation honestly and in good faith, they

---

35 A decision with reasons is known as a "speaking" determination. A decision without reasons is, logically enough, called a "non-speaking" determination.

36 It is convenient to say a little at this juncture about the distinction between speaking and non-speaking valuations or certificates, which to my mind is not a relevant distinction. Even speaking valuations may say much or little; they may be voluble or taciturn if not wholly dumb. The real question is whether it is possible to say from all the evidence which is properly before the court, and not only from the valuation or certificate itself, what the valuer or certifier has done and why he has done it.
(per Dillon LJ in *Jones and anor v Sherwood Computer Services plc* [1992] 1 WLR 277 (CA))

37 *Jones (M) v Jones (RR)* [1971] 1 WLR 840 (Ch).
38 [1991] 2 EGLR 103.
39 [1976] 1 WLR 403 (CA).

are bound by it. Even if he has made a mistake they are still bound by it. The reason is because they have agreed to be bound by it. If there were fraud or collusion, of course, it would be very different. Fraud or collusion unravels everything.

8.69 Dillon LJ has explained the steps that a court must take in considering an expert's decision:

> the first step must be to see what the parties have agreed to remit to the expert, this being, as Lord Denning MR said in *Campbell v Edwards* [1976] 1 W.L.R. 403, 407G, a matter of contract. The next step must be to see what the nature of the mistake was, if there is evidence to show that. If the mistake made was that the expert departed from his instructions in a material respect ... either party would be able to say that the certificate was not binding because the expert had not done what he was appointed to do.[40]

8.70 An expert who fails to produce a decision in accordance with its instructions is likely to have difficulty recovering its fees and will be open to a claim from the parties for having carried out the instructions negligently.

8.71 While the expert has to follow its instructions, it does not have to run the process as though it is a quasi-arbitration or adjudication. In *Bernhard Schulte GmbH & Co v Nile Holdings Ltd*,[41] Cooke J held:

> there is no is no requirement for the rules of natural justice or due process to be followed in an expert determination in order for that determination to be valid and binding between the parties.

8.72 Accordingly there is no requirement for the expert to hold hearings and invite submissions from the parties, and the expert is entitled (within the remit of the instructions) to use its own experience and knowledge in reaching its decision.[42]

8.73 This gives the expert considerable latitude. Even an expert who appears biased will not necessarily have its decision set aside by the court.[43]

8.74 It is vital, however, that the expert is not actually biased in favour of one of the parties. As the authorities above confirm, apart from a failure to comply with instructions, the only grounds for setting aside an expert's decision are fraud or collusion. This includes actual bias. Proving that an expert was actually biased is a very difficult task, requiring strong evidence to support it. This will not, however, stop a disappointed party from making such allegations. As a matter of good practice and self-preservation, therefore, a person performing an expert determination should seek to be as fair and balanced as possible as between the parties.

Vulnerabilities

8.75 It was noted above that there is no statutory immunity from suit for a professional carrying out an expert determination. Nor is there any protection at common law. It had been thought that a professional acting in a "quasi-judicial" role, such as an architect valuing an account, could not be sued for errors made in carrying out that function. This was

---

40 In *Jones and anor v Sherwood Computer Services plc* [1992] 1 WLR 277 (CA) at 287.
41 [2004] EWHC 977 (Comm).
42 The court will not imply terms of procedural fairness or natural justice into an expert determination agreement: *Owen Pell Ltd v Bindi (London) Ltd* [2008] EWHC 1420 (TCC).
43 *Bernhard Schulte GmbH & Co v Nile Holdings Ltd* [2004] EWHC 977 (Comm), helpfully summarised in *Owen Pell Ltd v Bindon (London) Ltd*, ibid.

emphatically dismissed by the House of Lords in *Sutcliffe v Thakrah*.[44] Subsequent courts have held *Sutcliffe* means that an expert making a decision in an expert determination has no such immunity either.[45]

8.76 Where does this leave the construction professional making an expert determination? If it follows its instructions and makes a negligent, but honest, error in reaching a decision, that decision will be enforced as between the parties to the dispute. The professional will, however, be at risk of a claim from the losing party for breach of its obligation of reasonable skill and care in carrying out the determination. Save in the most obvious cases, such a claim will not be without difficulties because it will not be easy to prove the breach of duty, given the limitations inherent in the expert determination process – in most cases the process is not supposed to be as exhaustive or as slow as arbitration or litigation.

---

44 [1974] AC 727 (HL).
45 See the discussion of Dillon LJ in *Jones and anor v Sherwood Computer Services plc* [1992] 1 WLR 277 (CA) at 283–288.

CHAPTER 9

Dispute resolution

Section A: informal dispute resolution	171
Section B: alternative dispute resolution	172
Section C: adjudication	174
Section D: the pre-action protocol	176
Section E: litigation	181
Section F: arbitration	184

SECTION A: INFORMAL DISPUTE RESOLUTION

9.1 Subject to obtaining the agreement of its professional indemnity insurer where required, there is no reason why a construction professional should not resolve a modest dispute which may raise possible professional negligence informally.

9.2 During the course of any project it is common for clients and the construction professionals they engage to differ over events that have occurred or costs that have been incurred. Such differences are generally resolved on the project or shortly afterwards in discussions. It is not uncommon for construction professionals to enter into discussions about their fees and charges when unpaid and for the client to use actual or potential errors on the part of the construction professional as a tool in negotiations. Plainly commercial life would be very difficult for any construction professional which felt prevented from making sensible agreements on small matters in order to prevent the emergence of larger dispute.

9.3 Many professional organisations require their members to provide a means for resolving complaints. Thus for example the ARB Code of Conduct provides:

> 10.1 You are expected to have a written procedure for the handling of complaints which will be in accordance with the Code and published guidance.
>
> 10.2 Complaints should be handled courteously and promptly at every stage, and as far as practicable in accordance with the following time scales: a) an acknowledgement within 10 working days from the receipt of a complaint; and b) a response addressing the issues raised in the initial letter of complaint within 30 working days from its receipt.
>
> 10.3 If appropriate, you should encourage alternative methods of dispute resolution, such as mediation or conciliation.[1]

---

1 The RIBA guidance to members of the public "It's Useful to Know" suggests that:

  i. If, unfortunately, things start to go wrong it is always best to try to talk to your architect before considering taking matters any further.
  Be prepared for your conversation:

9.4 Consequently there is no requirement for every dispute involving the possibility of an allegation of professional negligence to be treated as a potential claim requiring resolution by lawyers. However, particularly having regard to the usual requirements of professional indemnity policies, it is generally sensible for construction professionals to be cautious when seeking informal resolution of disputes.

9.5 In the first place, whilst the financial value of a dispute is a reasonably good guide to its importance, care must be taken to ensure that the construction professional knows precisely what is in issue. A problem may appear very minor but may in fact turn out to have substantial consequences. Insurers should always be informed of any potential claim which has the capacity to turn into something where the construction professional might have a need to look to its insurer.

9.6 Secondly, it will be a rare dispute where the construction professional will make a formal admission of liability. The better course is usually to offer some accommodation on fees as a commercial gesture in order to keep good relations. Thirdly, even the smallest accommodation should be adequately recorded in writing.

SECTION B: ALTERNATIVE DISPUTE RESOLUTION

9.7 Alternative Dispute Resolution, or "ADR", refers to a formal process by which two parties can resolve a dispute without involving litigation. For these purposes it is also used as an alternative to adjudication and arbitration, however it its wider coinage the term can apply to almost any system of compromising claims which exists outside of the courts.

9.8 There are a multiplicity of forms of ADR, but for the purposes of claims against construction professionals, mediation is almost universally the preferred route.

9.9 Mediation is a voluntary process under which parties engage a mediator to act as a facilitator of settlement. The mediator – a neutral person, often a lawyer or another construction professional – will spend a day or longer with the parties exploring the dispute and the options for settlement.

9.10 The entire process of mediation is protected by strict confidentiality so that nothing that transpires can be used in subsequent proceeding without the agreement of the parties. If the parties reach agreement the agreement itself can become a "public" document, although here too the parties have complete autonomy. As with all the other characteristics of the process, the way in which an agreement becomes binding and the use that may be made of it is regulated by the agreement which the parties have signed so such variation as they may agree. There is no limit on the number of parties to a mediation and indeed the process is particularly successful in comparison with other forms of dispute resolution when it comes to resolving multi-party disputes.

9.11 Some professional bodies expressly endorse mediation as a means of dealing with disputes against construction professionals,[2] but it is very widespread as a tool by which

---

- check your appointment agreement (i.e. the terms of engagement with the architect) which you should have agreed together at the outset. These should have been set out in writing by the architect.
- Remind yourself of the specific services he or she (or the practice) was contracted to provide, plus any particular requirements that were part of your agreed brief (such as the project budget or timetable)
- Identify where you believe the requirements of your agreement have not been met.

2 The RIBA operates a fixed fee mediation scheme for smaller claims.

claimants and construction professionals and their insurers can resolve claims. Almost every large claim which reaches trial will have first passed through the mediation process.

9.12 Whilst there is no set process for mediation, the vast bulk of mediations involving construction professionals will follow the same pattern.

9.13 The parties must agree to go to mediation. As set out below, the courts favour mediation and can penalise litigants who unreasonably refuse to attend mediation. Arbitrators may take a similar approach. It is therefore rare for parties involved in claims against construction professionals to refuse to mediate.

9.14 Rather, the usual difficulty over agreeing to mediate is not whether to do so, but when and in what circumstances. For mediation to succeed the parties often need to know quite a lot about each other's cases. This may mean that a mediation at an early stage in the life of a dispute stands less chance of success than it would at a later stage. For example, one party may insist on seeing documents which will only become available on disclosure (see below). Often the involvement of third parties dictates the timing of mediations. A third party which is not part of legal proceedings, but against which a written claim has been made, may contend that it should not be part of a mediation between two parties who are in legal proceedings. One of the other parties may refuse to mediate until that third party is present. Because mediation is a voluntary process there is no effective mechanism for preventing one party putting obstacles in the way which, although they may not prevent mediation occurring at some stage, may well delay it.

9.15 Once the parties have agreed to mediate they must agree a mediator. Mediators for disputes involving construction professionals are normally drawn from a small group of persons regularly holding themselves out as mediators in construction and commercial litigation, and are usually lawyers or construction professionals. Generally they are accredited by one of the recognised providers of mediation services. Although they will usually be well-known to the lawyers who act this field, they will be independent of the parties. It is the mediator who usually provides the parties with the mediation agreement which sets out the terms under which they will conduct the mediation.

9.16 In advance of the mediation the parties will usually agree a bundle of essential documents for use in the mediation and will exchange position papers. Position papers are encouraged to be positive and constructive, although it is common for them merely to set out the parties' cases.

9.17 The standard mediation will last a day and will take place at hired rooms in an institute, solicitors' offices, hotel or other suitable space. The parties will be represented by lawyers, but it is usual for at least one person from the construction professional and one person from the professional indemnity insurer to attend.[3] Each party and the mediator will have their own rooms. The mediation may well commence the day with a "plenary session" at which the parties are convened in the same room and have the opportunity to make short opening statements.

9.18 Thereafter, the mediator will shuttle between the parties, both taking representations made by each as to each other's case and discussing areas of common ground. A skilled mediator will attempt to build consensus on such issues as can be agreed and to identify for all parties the risks which need to be taken into consideration if the matter is

---

3 It is usually a term of the mediation agreement that it is attended by persons with authority to settle.

litigated or arbitrated. In this way the mediator hopes to encourage the parties to move from a process of dialogue to a process of bargaining, resulting eventually in settlement.

9.19 Whilst many mediations result in settlement being achieved on the day, a substantial proportion achieve constructive results short of settlement. Even if the parties do not come sufficiently close to resolve the dispute at the mediation, the process may enable them to reach that point in the weeks that follow. It is a rare mediation which results in the parties being in either the same or a worse position with regard to avoiding trial. For this reason it is important for parties to disputes involving construction professionals, whatever their views on the merits, to look positively upon the process rather than to treat it a step which must be complied with in order to avoid potential costs sanctions.

SECTION C: ADJUDICATION

9.20 Most construction professionals will be familiar with adjudication. In the United Kingdom it is a process of dispute resolution introduced by Parliament[4] and intended to provide speedy interim decisions in relation to (mostly) disputes about payment. Parties to "construction contracts" must include provisions for adjudication in the contract and in default of compliant provisions a statutory scheme will be implied.

9.21 The legislative intent is to provide a means whereby disputes could be resolved by an impartial adjudicator, generally within 28 days, whose decision would be binding upon the parties until overturned (or confirmed) in court or arbitration.

9.22 The genesis of the scheme was concern that larger construction companies were exploiting smaller ones by delaying payments due (or in some cases simply not paying at all), taking advantage of the fact that there was no speedy and cheap means by which they could be made to meet their obligations. It seems unlikely that Parliament had disputes involving construction professionals uppermost in its mind when it passed the Act. However, its provisions are very broad and the definition of "construction contract" and "construction operations" are sufficiently widely drawn to encompass almost all contracts between construction professionals and their clients, including collateral warranties.[5]

9.23 Construction professionals who act at a remove from the works themselves may fall outside the Act (for example the provision of expert or other evidence)[6] but the provision of services in relation to ongoing construction works will almost always fall within it. The only exception is contracts with the occupiers of residential dwellings (private houses or flats), which are excluded from the scheme.[7]

---

4 Part II of the Housing Grants, Construction and Regeneration Act 1996.

5 Section 104(1) provides that a construction contract is an agreement for the carrying out of "construction operations" and arranging for such operations to be carried out by others. Section 104(2) includes within the definition any agreement for the provision of advisory services whether architectural, design, surveying, building, engineering, interior or exterior decoration and landscaping. Construction operations is very broadly defined and applies to almost all construction projects, with specific exceptions (for example, steelworks). A professional engagement to administer a building contract has been held to come within the section, as has a collateral warranty offered by a contract administrator: see *Parkwood Leisure Ltd v Laing O'Rourke Wales & West Ltd* [2013] BLR 589.

6 *Fence Gate Ltd v James R Knowles Ltd* (2001) 84 Con LR 206 (TCC).

7 Section 106. It is possible for such persons to agree *contractual* obligations to adjudication (so that, for example, a construction professional could adjudicate for its fees), but such provisions are at risk of being struck down as unreasonable under the Consumer Act 2015.

9.24 A party to an adjudication agreement may seek to adjudicate a "dispute" at any time.[8] This means that provided a dispute has arisen (and dispute is given a broad meaning to include any difference) a notice of adjudication may be served. Thus, if there is a disagreement between the engineer and the client as to the likely additional costs of further works which the engineer has advised the client does not have to wait until the end of the contract or even the incurrence of the costs. It can seek the decision of an adjudicator from the moment a dispute arises. There is no requirement for any kind of exchange of views or attempt to resolve the matter. This also means that a notice of adjudication can be served years after works have completed.

9.25 Adjudication does not remove the parties' substantive rights and so a construction professional's statutory and contractual limitation rights are preserved, but it does mean that a client can wait many years before commencing an adjudication.

9.26 Once a notice of adjudication is served the timetable for reaching a decision is extremely tight. Unless the parties agree the adjudicator must provide a decision within 28 days of the notice, although the adjudicator has power to extend time for a further 14 days with the agreement of the referring party.[9] A tight timetable is put in place for a response and the submission of written evidence. There is no provision for any kind of disclosure and it is rare for the adjudicator to hear the parties in person. The issues are decided on paper. In practice this can mean that the responding party has little time to prepare and present its case and may often be ambushed. The more complex the dispute (and many professional negligence disputes are necessarily complex) the less appropriate adjudication appears as a means of resolving the parties differences.[10]

9.27 An adjudicator's judgment is binding (between the parties) until the relevant dispute has been determined in court or in arbitration (or by agreement).[11] Thus if an employer obtains an adjudication award against an architect for £500,000 that sum becomes a debt which the architect (or, realistically, its insurer) must pay to the employer. It then falls to the architect to commence proceedings to recover that sum, whether in litigation or arbitration.[12] Plainly this can provide the adjudicating party with a considerable tactical advantage, even though the adjudicator's decision carries no weight in any subsequent proceedings.

9.28 Of course, just as an employer may commence an adjudication against a construction professional, a construction professional may commence an adjudication against an employer seeking its unpaid fees and costs. If an employer unsuccessfully defends such a claim on the basis of alleged breaches of duty the employer will be indebted to the construction professional and will have to commence legal proceedings in order to recover its monies.[13]

9.29 Adjudication awards can seldom be challenged in the courts. The scheme is intended to provide for quick interim justice and its entire point would be undermined if

---

8 Section 108.
9 Section 108(2).
10 There are occasional comments in the authorities about the fact that adjudication is not well suited to professional negligence claims – see for example the Scottish case of *Whyte and Mackay Ltd v Blyth & Blyth Consulting Engineers Ltd* [2013] CSOH 54 – but there is no rule of law which prevents any dispute falling within the definitions of the Act being adjudicated.
11 Section 108(3).
12 *Aspect Contracts (Asbestos) Ltd v Higgins Construction Ltd* [2015] UKSC 38.
13 The matter cannot be re-adjudicated. An adjudicator's decision binds subsequent adjudicators.

decisions were capable of appeal or other challenge save on very limited grounds. A mistaken decision cannot be challenged merely because it is mistaken (even if the mistake is plain and obvious).[14]

9.30 Decisions have been successfully challenged where adjudicators have strayed outside their jurisdiction (for example if the decision fails to decide the dispute referred)[15] and where the adjudicator has failed to comply with the rules of natural justice (for example failing to give one party a proper chance to address an important issue).[16] In exceptional cases, where there is a real risk that the applicant may become insolvent, an unsuccessful respondent may be able to obtain a stay of execution.[17] Reference should be made to the specialist works dealing with the law and practice of adjudication.[18]

9.31 Although employers do commence adjudications against construction professionals, the use of this tactic remains uncommon. Rightly or wrongly the advisers of employers often take the view that, save in very plain cases, adjudicators will be reluctant to rule on matters which are not only complex but often decided by reference to expert evidence. It is often thought that adjudication will merely add a further layer of cost and delay to proceedings[19] and indeed may backfire as professional indemnity insurers have the deep pockets necessary to commence litigation or arbitration to claw back sums which (in their view) should not have been paid.[20]

SECTION D: THE PRE-ACTION PROTOCOL

9.32 In the United Kingdom the courts encourage prospective parties to litigation to explore those claims in correspondence before proceedings are issued. The process aimed at facilitating settlement and narrowing areas of dispute so that if proceedings are issued costs and time are not wasted dealing with irrelevant or unnecessary issues.

9.33 Court encouragement of the process has led to the development of formal protocols to guide the pre-action process. Whilst there is a separate protocol for professional negligence claims, disputes involving construction professionals fall to be dealt with under the Pre-Action Protocol for Construction and Engineering Disputes.[21]

9.34 Failure to follow the Protocol without good reason may result in adverse award of costs in subsequent court proceedings. Whilst the use of this Protocol is not generally

---

14 There are a large number of cases in which the courts have declined to interfere in the adjudication process. For a prominent explanation see *Carillion Construction Ltd v Devonport Royal Dockyard Ltd* [2005] EWCA Civ 1358.

15 See, for example, *Herbosch-Kier Marine Contractors Ltd v Dover Harbour Board* [2012] EWHC 84 (TCC).

16 See, for example, *ABB Ltd v Bam Nuttall Ltd* [2013] EWHC 1983 (TCC).

17 See *Wimbledon Construction Co 2000 Ltd v Vago* [2005] EWHC 1086 (TCC).

18 For example, Sir Peter Coulson, *Coulson on Construction Adjudication* (Third Edition).

19 Cost is relevant because, unlike in litigation, the parties' costs in an adjudication are not recoverable. The only "costs" that the Adjudicator can award are his own fees. Delay is relevant not so much because the process itself takes time (it can be completed within 28 days) but because the claimant and its lawyers will generally be tied up for weeks if not months in preparation.

20 Indeed, claimants tend to commence adjudications against construction professionals either where there is a very clear case and those acting for the construction professional are perceived as being unreasonable in refusing to accept this or where there is substantial doubt as to whether the construction professional has insurance cover for the relevant claim and the claimant believes that a successful adjudication will flush this out.

21 Now in its Second Edition: www.justice.gov.uk/courts/procedure-rules/civil/protocol/prot_ced

mandated by either arbitration agreements or the rules of procedure commonly adopted for arbitrations, arbitrators commonly have regard to whether parties have followed an appropriate pre-action process and, as a generality, it can be assumed that compliance with the Protocol is as much a condition precedent to arbitration as it is to litigation.

9.35 The Protocol requires that before commencing proceedings the claimant should send the proposed defendant stating, amongst other matters, a brief summary of the claim or claims including (a) a list of principal contractual or statutory provisions relied upon (b) a summary of the relief claimed including, where applicable, the monetary value of any claim or claims with a proportionate level of breakdown. The extent of the brief summary should be proportionate to the claim. Generally it is not expected or required that expert reports should be provided but, in cases where they are succinct and central to the claim they can form a helpful way of explaining the Claimant's position.

9.36 Because the Protocol is applicable generally to construction disputes this wording is very broad. In the usual case involving a construction professional, the Pre-Action Protocol Letter of claim will be neither the first intimation of a claim,[22] nor will it be a brief summary: save in the most straightforward cases professional negligence claims against construction professionals carry a degree of complexity and it is commonly expected that that the letter of claim will set out in some detail (1) the essential facts; (2) the duties alleged to have been owed by the construction professional; (3) a reasonably full explanation of how and in what ways the construction professional is said to have acted in breach of those duties; (4) a reasonably full account of how and in what ways it is said that those breaches of duty caused the claimant loss; and (5) a reasonably detailed breakdown of the loss which is said to have been caused.

9.37 The letter will commonly annex copies of essential documents (for example, the engagement) and, although it is not common for experts reports to be provided, the allegations of breach of duty are expected to be advanced with the support of an expert in the relevant field, who will be identified in the letter.

9.38 It follows that for the purpose of bringing claims against construction professionals the Pre-Action Protocol letter of claim will usually contain at least as much detail as will be included in the Particulars of Claim if proceedings are commenced. Plainly this means that Pre-Action Protocol letters of claim are generally expensive involving considerable lawyer and expert input. Whilst on its face this might seem to run contrary to the objectives of the Protocol (the resolution of disputes at minimal cost), in practice a detailed and well thought out Letter of Claim saves to save both costs and time in the long run. It enables the Claimant to set out its full case in a way which tells the construction professional precisely what it is that must be responded to and it requires a response of appropriately equivalent detail and care so that, at an early stage, the parties are able to adequately to evaluate their positions and consider compromise as an alternative to litigation.

9.39 The Protocol provides that the construction professional should provide an early acknowledgement of the Pre-Action Protocol Letter of Claim (within 14 days of receipt), providing the name of its insurer and (within 28 days) taking any jurisdictional objection (for example that the parties have agreed arbitration as the means of resolving their disputes or that the Claimant proposes to sue the wrong defendant).

---

22 It is common for a prior short letter to be sent warning of the possibility of a claim, if only for the purpose of ensuring that the insurers of the construction professional are on notice (thereby minimising the risk of the insurer declining cover and delay) and in order to request documents.

9.40 Absent such preliminary objection, the construction professional is required to provide the Claimant with a letter of response and to do so within 28 days. The letter of response should contain, amongst other information, "a brief and proportionate summary of the Defendant's response to the claim or claims and, if the defendant intends to make a Counterclaim, a brief summary of the Counterclaim".

9.41 In practice it is correspondingly rare for a construction professional to provide either a brief summary of its response or to do within 28 days. Just as the Letter of Claim will set out the Claimant's case in some detail so the Letter of Response (prepared by the lawyers instructed by the construction professional's indemnity insurers) will generally be expected to set out a full defence to the proposed claim, taking issue with the individual components of the claim and explaining why these are not accepted (where that is the case) in a way which allows the Claimant to understand the case that the construction professional is making. On its own this would invariably require more than 28 days.

9.42 However, the reality of claims against construction professionals is that whilst insurers may engage lawyers on receipt of an intimation of a claim, those lawyers are generally not asked to investigate the claim until the Pre-Action Protocol Letter arrives. It is probably at that point that the lawyers will begin gathering relevant documentation, interviewing the construction professional and engaging an expert. Whilst there is no rule of thumb for the time necessary for a proper Letter of Response it is not uncommon for the parties to agree that a construction professional should have 6 to 8 weeks to put together its response (8 weeks being the maximum period permissible).

9.43 Again, whilst on its face this appears to be a contravention of the Protocol, it is in fact a proper application of the guidance. Provided that the construction professional responds constructively, with a sufficiently full and detailed Letter of Response, the parties will be in a much better position to assess whether they can avoid the costs and uncertainties of litigation (and to assess which issues are really necessary for resolution by litigation) than they would be if the exercise had been rushed by sticking to the 28 day timetable and confining the content of the Letters to a brief summary.

9.44 The Protocol does not require a Claimant to respond to the Defendant's Letter of Response unless the Defendant makes a Counterclaim. However, it is common and (it is suggested) good practice for some kind of response to be provided. The content of any such response will depend upon the precise nature of the issues in dispute and the positions taken by the parties. In a particular case it may be appropriate simply to note the Defendant's position but explain (briefly) why the Claimant is unpersuaded by it.

9.45 A more detailed response is advisable where either (1) the Claimant accepts part of the Defendant's arguments, but none the less maintains its claim (the Claimant needs to explain why so that the issues are narrowed) and (2) where the Claimant wishes to raise different facts and matters to those raised in the Letter of Claim as part of its explanation as to why the Defendant's arguments are not correct.

9.46 The Protocol recommends a third stage which is a meeting between the parties. Paragraph 9.1 states:

> Within 21 days after receipt by the Claimant of the Defendant's letter of response, or (if the Claimant intends to respond to the Counterclaim) after receipt by the Defendant of the Claimant's letter of response to the Counterclaim, the parties should normally meet.

The words "should normally" indicate that the Protocol allows of exceptions. In rare cases there may be no merit in a meeting at all. In the more usual case, the parties will agree to meet but will further agree that a meeting cannot sensibly be arranged within 21 days. This is particularly the case if the parties intend to bring experts and other persons to the meeting.[23]

9.47 Whilst the guidance is careful not to be prescriptive the purpose of the meeting is explained in these terms:

> Generally, the aim of the meeting is for the parties to agree what are the main issues in the case, to identify the root cause of disagreement, and to consider (i) whether, and if so how, the case might be resolved without recourse to litigation, and (iii) if litigation is unavoidable, what steps should be taken to ensure that it is conducted in accordance with the overriding objective as defined in rule 1.1 of the Civil Procedure Rules. Alternatively, the meeting can itself take the form of an ADR process such as mediation.

9.48 Paragraph 9.4 contains a checklist of issues which the parties should consider if litigation appears unavoidable:

9.4.1 if there is any area where expert evidence is likely to be required, how expert evidence is to be dealt with including whether a joint expert might be appointed, and if so, who that should be; and (so far as is practicable);

9.4.2 the extent of disclosure of documents with a view to saving costs and to the use of the e-disclosure protocol; and

9.4.3 the conduct of the litigation with the aim of minimising cost and delay.

9.49 The contents of the meeting are "without prejudice". That means that they cannot be referred to by either side in any subsequent proceedings. However, the Court will be permitted to know whether the parties reached any agreements (and what they were) and, for the purposes of possible cost penalties, to know whether any party refused to attend a meeting and if so why.

9.50 The Protocol process is deemed to be complete "at the completion of the pre-action meeting or, if no meeting takes place, 14 days after the expiry of the period in which the meeting should otherwise have taken place". However, it does not follow that because the Protocol process has come to an end the parties must proceed to litigation or arbitration. The logic of the Protocol is save where litigation or arbitration is unavoidable, continued efforts should be made to find a resolution. There is no *requirement* for either party to continue exploring these options or to continue negotiating, but it may be in both parties' commercial interests for this to happen.

9.51 In this context regard should be had to a further, very important, consideration which is frequent in disputes involving construction professionals. Claims against architects, engineers, surveyors and other construction professionals will rarely involve the making of counterclaims (save for outstanding fees) but they will often involve widening the dispute to bring in other persons (for example other construction professionals). For the reasons which have been given elsewhere, it is common for losses caused to clients on construction projects to have been the product of a series of acts or omissions in which a number of different persons were involved.

---

23 Paragraph 10.1 provides that: "The parties may agree longer periods of time for compliance with any of the steps described above save that no extension in respect of any step shall exceed 28 days in the aggregate".

9.52 If a claim is made against the architect for the consequences of delay caused by inadequate design, the architect may wish to contend that the person responsible, or partly responsible, was the engineer. Guided by its lawyers the architect (or rather the architect's insurer) may want to bring the engineer (or rather the engineer's insurer) into the dispute with the Claimant so that the architect can obtain a contribution. That will generally necessitate the architect sending its own Pre-Action Letter of Claim to the Engineer. This necessarily builds delay into the overall process of seeking to resolve the claim short of litigation even if it does not affect the strict timetable under the Protocol.

9.53 As a generality, claimants may find it serves their interests to be tolerant of reasonable delays. Disputes are often harder to settle when important parties are not involved in the negotiations. Even if the claimant is firmly of the view that it has no interest in the construction professional's claim against a third party, it is often prudent to allow time for this claim to be intimated and responded to.

9.54 Although the primary intention of the Protocol is to allow parties the fullest opportunity to resolve their differences, the Protocol does not itself prescribe mechanisms by which settlement can be achieved. If the parties have come sufficiently close (particularly on a straightforward claim) resolution of the dispute can be achieved by a process of offer and counter offer by the parties solicitors under cover of "without prejudice" protection. Typically this is either entirely conducted in writing, or agreement reached by oral offer and acceptance is recorded in letters passing between the solicitors. However, where parties are sufficiently close in their appreciation of the merits that settlement becomes a realistic prospect, the more common route to resolution following the Protocol is for the parties to agree to a mediation (or, unusually, some other form of structured alternative dispute resolution).

9.55 Whilst the Protocol should always be followed unless there is good reason not to do so, there will be occasions where such good reason exists. The most common are cases where a limitation date is approaching (or is arguably approaching) or where litigation has commenced between two other parties, one of which then decides to bring in the construction professional by third-party proceedings (see below).

9.56 Here the spirit of the Protocol may guide the proper conduct of the parties. Thus the fact that limitation is approaching does not, of itself, mean that the claimant has to issue protective proceedings. If there is sufficient time, the better course is for the parties to agree a "standstill agreement", the effect of which is to stop time running for limitation purposes so that the parties can follow the Protocol process.[24] If there is insufficient time to allow for the negotiation of a standstill agreement it may be appropriate for the claimant to issue a claim form, but for the parties then to agree a "stay" or to agree a "stay" after the service

---

24 It must be kept in mind that "standstill agreements", like any contract, only serve the parties' interests if they are well drafted. Problems with poorly drafted standstill agreements have led some Judges to express the view that the safe and better course is to commence proceedings and then agree a stay. In *Russell v Stone* [2017] EWHC 1555 (TCC), Coulson J observed:

> If limitation is an issue, and the claim needs further work, or the Pre-Action Protocol process has not been activated or completed, the TCC Guide is very clear: para.2.3.2 states that the claimant can commence proceedings and then seek a stay of, say, six months, in order to follow and complete the Protocol process.

However, it is respectfully suggested that, save in very plain cases, a standstill agreement is still preferable to the issue of proceedings. It is generally preferable to commence proceedings when it is clear what is in issue and why.

of the initial statements of case. In either case the purpose would be to permit the parties to follow the Protocol before further litigation commences.

9.57 The joinder of a construction professional to existing litigation will generally be more difficult to reconcile with the Protocol (the existing parties and probably also the Court will have limited appetite for delays), but that does not mean that, in an appropriate case, the directions in the third-party proceedings cannot be staggered to allow the parties to narrow their differences as much as possible before the litigation reaches an advanced stage.

9.58 Finally, whilst the Court will not police the Protocol process, it will not be slow to penalise parties who abuse it. Paragraph 5 states:

> The overriding objective (CPR rule 1.1) applies to the pre-action period. The Protocol must not be used as a tactical device to secure advantage for one party or to generate unnecessary costs. In many cases, including those of modest value, the letter of claim and response can be simple and the costs of both sides should be kept to a modest level. In all cases, the costs incurred at the Protocol stage should be proportionate to the complexity of the case and the amount of money which is at stake. The Protocol is not intended to impose a requirement on the parties to marshal and disclose all the supporting details and evidence that may ultimately be required if the case proceeds to litigation.

SECTION E: LITIGATION

9.59 The primary means of taking legal action against construction professionals in the United Kingdom remains litigation. Although compulsory arbitration clauses are common, most engagements omit them.[25] In any event, as has been set out above, many actions against construction professionals are contribution proceedings where some other person (for example a contractor or another construction professional) seeks to bring the construction professional within existing proceedings. If disputes have not been resolved by the pre-action protocol process, they will generally proceed to court.

9.60 The choice of court is largely determined by the value of the claim. The following guidance applies to cases in England and Wales.[26] Under the current rules, claims should only be commenced in the High Court where the value exceeds £100,000. This is not an inflexible guide and the true position as to claim allocation is more nuanced. Professional negligence cases against construction professionals can be commenced in the High Court if their value is less than £100,000 provided that the level of complexity justifies that course.

9.61 The natural home for such cases is the Technology and Construction Court ("the TCC"). The TCC mainly sits in London, but there are also branches of the TCC in many of the major regional cities. As a general rule cases worth less than £250,000 should be dealt with outside the TCC in London and (for lower value cases) in the county court centres which also offer technology and construction court expertise.[27] Again, however, cases of a lesser value may merit being heard in London if they are sufficiently complex. There is no requirement to try high-value cases in London. These can be dealt with by the TCC in the regional centres (and that may be appropriate if all the parties are based nearer to that centre).

---

25 For the reasons which are set out below.
26 The courts in Northern Ireland and Scotland have separate rules.
27 *West Country Renovations Ltd v McDowell* [2012] EWHC 307.

9.62 The judges of the TCC not only have specialist expertise in construction matters, but they also impose a high degree of case management upon the cases in their court. Construction professionals can therefore expect cases involving them not only to proceed at as brisk a pace as can reasonably be managed but also to have a considerable degree of scrutiny at an early stage with the intention of restricting the proceedings to what is really necessary. A typical timeline might be as follows:

- January: the claimant serves its claim form and Particulars of Claim on the solicitors who have agreed to act for the construction professional and its professional indemnity insurer
- February: those acting for the construction professional serve its Defence and possibly a Request for Further Information. At this stage the construction professional might also commence third-party proceedings against another person, seeking to bring that person into the action.
- March: the claimant serves its Reply and any response to a request for further information. If proceedings have been taken against another person it is possible that that person's defence might be served at this time.
- May: the first Case Management Conference occurs. This is a hearing (in front of the Judge in the TCC) when the court decides what directions to give to the parties to take the case to trial. The court will generally decide how long the trial needs to be and will fix a date. Direction will include permission to call expert evidence. The court will usually scrutinise the parties' estimates of what legal costs they are likely to incur and may make restrictions on the costs which will be recoverable if a party is successful.
- July: the parties provide disclosure. They undertake a search of the documentation (including electronic documentation) in their possession and control and identify those documents which are material to the issues in the case. These are then made available to the other parties in the proceedings to consider.
- September: the parties exchange witness statements. Save in very rare cases, in civil proceedings in England and Wales the non-expert evidence that a party intends to give at trial is contained in a witness statement so that all parties can know what evidence is going to be given long before they get to trial.
- November: the parties' experts' exchange their reports. By this time the parties' experts will have met to try to narrow down their differences. After the service of reports they will meet again to prepare a statement of agreement and disagreement. Sometimes it will be necessary for there to be further reports.
- January: the parties will attend at a "Pre Trial Review". This is hearing usually listed at least a month before the trial date which allows the court to check that the case is ready for trial and to give any directions which may be necessary.
- March: Trial: trials of actions against construction professionals generally follow the same pattern – the court will deal briefly with opening legal statements and arguments from the lawyers; the claimant will call its witnesses of fact, who will rarely be allowed to add to their witness statements but whose oral evidence will chiefly take the form of cross-examination; the legal team acting for the construction professional will then call its witnesses of fact (typically the persons involved on the project which is the subject of the action), each of whom will be similarly

cross-examined; the court will then receive evidence from the parties experts, who are usually called in accordance with their discipline, so that the liability experts are called sequentially, followed by the quantum experts; as with the witnesses of fact, the reports of the experts will stand as their evidence and their oral evidence will be mostly obtained by cross-examination; after the conclusion of the expert evidence the court generally hears submissions from the lawyers, which are often both written and oral. In rare cases the Judge will be able to provide a judgment right away. In the usual case, Judgment will be "reserved".

- May: the Judge hands down Judgment. The parties will usually obtain a draft of the Judgment about a week in advance of the date when Judgment is to be handed down. It is confidential at this stage, but it permits the parties to consider matters such as which party is going to pay what legal costs and whether the losing party wishes to apply for permission to appeal. At the handing down of Judgment these matters are sometimes agreed (so that nobody needs to attend) or are the only matters to be dealt with by the court.

9.63 If a construction professional is found liable to pay damages and/or legal costs responsibility for making that payment will generally fall it its insurer. In a case where a construction professional recovers its fees (either because it is the claimant, or under a counterclaim) the person liable to pay the fees will generally have between 14 and 28 days to do so.

9.64 This timeline is appropriate for a modest value case between an employer and a construction professional. More complex claims, involving more parties, can be much more difficult to bring to trial and can take considerably longer. Moreover during the course of the proceedings one party may bring an application to the court which has a dramatic effect upon the overall merits (for example, an application to obtain judgment on the whole or part of a claim on the basis that it is bound to fail).

9.65 The vast majority of claims against construction professionals commenced in the TCC are resolved before trial (and sometimes they are resolved during trial). It is relatively rare for cases to be fought to Judgment. Cases can be resolved at any stage in the proceedings, but typically there are two complementary mechanisms which bring this about.

9.66 Legal costs can be substantial for cases against construction professional and the general rule in all substantial court proceedings in the United Kingdom is that the loser pays the other party's costs.

9.67 Under the rules devised for court proceedings in England and Wales parties may make offers to each other to settle the proceedings on terms. These offers are kept confidential from the court until after Judgment. If a party does worse than the offer, it will generally be penalised in terms of the costs it has to pay. Thus, at an early stage in the proceedings those acting for an architect may offer to pay the claimant £250,000 plus its costs. If that offer is rejected and the claimant only recovers £240,000 at Judgment, the claimant will generally be ordered to pay the construction professional its legal costs of the proceedings from the date when the offer should have been accepted. A claimant may make an offer which, if the construction professional does worse, will enable the claimant to claim a higher rate of costs and interest.

9.68 Typically the conduct of claims against construction professionals is marked by an exchange of such offers (and there is no limit on how many can be made) which can bring the parties sufficiently close so that they settle.

9.69 The other settlement mechanism is mediation. Whilst courts do not attempt to regulate the parties' attempts to mediate, they do expect the parties to attempt mediation unless there is a very clear reason why it would be a waste of time. The practice in the TCC in London is not to build a specific window for mediation into the directions timetable, but the desirability of such a window may be a persuasive factor in any argument that a particular period (for example, the period between the end of disclosure and the exchange of witness statements) should be slightly longer.

9.70 Mediations during the course of proceedings are often more successful than mediations occurring before proceedings have started. The parties will have set out their cases in more forensic detail and (something which is often critical to actions against construction professionals) they will both have seen each other's contemporaneous documentation. The costs risks of carrying on will be both more apparent and more imminent. There will be less scope for hoping that the other party will lose interest or that something will turn up.

SECTION F: ARBITRATION

9.71 Arbitration is a process of dispute resolution whereby the parties agree between themselves as to the appointment and jurisdiction of the tribunal which decides their dispute.[28] It is similar to litigation in terms of its process, but unlike litigation it is private and paid for by the parties. In the commercial world parties choose arbitration chiefly because of this confidentiality and, to a lesser extent, because it can be quicker and more final (that is, less vulnerable to appeal) than litigation.

9.72 In order for a dispute involving a construction professional to be referred to arbitration the parties must have entered into an arbitration agreement. Many standard form engagements for construction professionals include such provisions, which are designed to require disputes between construction professionals and their clients to be arbitrated.

9.73 Most employers (and particularly sophisticated employers) avoid using such provisions and will not agree to arbitration provisions in bespoke engagements. This is because they are generally perceived to favour the construction professional. In part this is because confidentiality is often thought to be more important to the construction professional than to the employer, but in reality it is because if something goes wrong on a modern construction project there is quite a high chance that the employer will want to be able to take proceedings against more than one person. In court proceedings the employer can sue both the contractor and the architect for the same loss and leave them to blame each other. Absent very unusual circumstances, arbitration does not allow this. It is a purely bilateral form of dispute resolution.

9.74 The Arbitrator will usually be a lawyer or a construction professional experienced in construction disputes. It is rare for arbitration agreements in the engagements of construction professionals to name a particular arbitrator and the usual arrangement is that the parties either agree an arbitrator or agree that some competent institution, for example the RICS, will nominate an arbitrator.

---

28 In *O'Callaghan v Coral Racing Ltd* [1998] All ER (D) 607 (CA), Hirst LJ said:

> the hallmark of the arbitration process is that it is a procedure to determine the legal rights and obligations of the parties judicially, with binding effect, which is enforceable in law, thus reflecting in private proceedings the role of a civil court of law.

9.75 The parties may have agreed to a particular scheme of arbitration rules covering matters such as evidence and procedure or the arbitrator may have freedom to adopt a particular set of rules. The stages of an arbitration are not dissimilar to the stages of litigation as set out above, although the process may be slightly less formal and there is usually more flexibility in the timetable.

9.76 The Arbitrator will produce a judgment in the form of an "award", but because of confidentiality this is not make public. Unlike litigation, there is no process of appeal. Arbitration awards can be challenged in very narrow circumstances, but the process is intended to provide finality. This in itself may make it unsuitable for the disposal of claims against construction professionals which involve complex points of law.

9.77 Because arbitration is confidential, there is little other than anecdotal evidence to suggest the volume of claims against construction professionals which are arbitrated. That anecdotal evidence suggests that litigation is much more prevalent.

INDEX

ACA *see* Association of Consultant Architects
ACE *see* Association of Consulting Engineers
adjudication 9.20
   see also adjudicators
adjudicators
   appointment 8.48
   immunity from suit 8.57
   obligations 8.53
   see also adjudication
ADR *see* alternative dispute resolution
alternative dispute resolution 9.7
appointment *see* contract
apportionment
   and contribution 6.106
arbitration 9.71
   see also arbitrators
arbitrators
   appointment 8.31
   immunity from suit 8.42
   obligations 8.34
   see also arbitration
architect
   administration of the building contract 5.119
   Architects' Registration Board and 2.7
   contracts with 2.10
   design obligations 5.40
   duties to third parties 3.55, 3.87, 3.91
   estimates 5.23
   examination of the site 5.4, 5.11
   function 2.3
   inspection 5.149
   and insurance 7.1
   meaning 2.1
   planning permissions and 5.13
   reasonable skill and care 2.15, 4.6
   responsibility for the work of others 5.81, 5.86
   Royal Institute of British Architects and 2.2, 2.7
   specialist expertise 4.46
   Standard Form of Agreement for the Appointment of an Architect 2.9
   warranties 2.17

ARB *see* Architects' Registration Board
Architects' Registration Board 2.7
Association of Consultant Architects 2.7
Association of Consulting Engineers 2.26
   conditions of engagement 2.27
Association of Project Management 2.53

betterment 6.51
breach of duty *see* duty of care
building surveyors
   meaning 2.55
   function 2.56
   contracts 2.58

causation
   break in chain of causation 6.32
   and damages 6.43
   discussion 6.6, 6.25
   hypothetical actions and 6.35
   remoteness 6.16
   scope of duty 6.7
certification
   under building contract 5.133
contract
   architects 2.10
   collateral warranty and 3.5, 3.108
   concurrency of contractual and tortious duty 3.77
   duties owed to persons other than original client 3.87
   formation of 3.7
   engineers 2.27
   limitation 3.34, 3.35
   net contribution clause 3.36
   novation and 3.27
   quantity surveyors 2.41
   variation 3.15
contribution 6.99
   and apportionment 6.106
   net contribution clause 3.36
contributory negligence 6.90

INDEX

construction professional
  meaning 1.3
copyright
  disputes 3.152

damages
  consequential losses 6.52
  costs of rectification and 6.43
  diminution in value and 6.43
  liability to third parties 6.58
  mitigation and 6.70
  personal injury, inconvenience and distress 6.62
  principles behind 6.43
  professional fees 6.66
  remoteness and 6.16
  scope of duty and 6.7
  settlement and 6.60
  wasted expenditure 6.52
Defective Premises Act 1972
  application 3.113
definitions
  adjudicator 8.45
  arbitrator 8.28
  architect 2.1
  engineer 2.19
  expert determination 8.59
  expert witnesses 8.4
  professional 1.3
  project manager 2.46
  quantity surveyor 2.31
design
  obligations of architects and engineers 5.40
  co-ordination and integration 5.71
  novel design 5.77
duty of care
  architects' and engineers' duties to third parties 3.55, 3.87, 3.91
  duty to review design 4.64
  economic loss 3.57, 3.76
  reasonable skill and care 4.1
  third parties and 3.87
  tort of negligence 3.45
  standard of care 4.21
duty of care deed *see* contract: collateral warranty

engineer
  ACE Conditions of Engagement 2.27
  administration of the building contract 5.119
  Association of Consulting Engineers 2.26
  contracts 2.27
  design obligations 5.40
  duties to third parties 3.55, 3.87, 3.91

Engineering Council 2.26
estimates 5.23
examination of the site 5.4, 5.11
function 2.22
meaning 2.19
inspection 5.149
Institution of Civil Engineers 2.26
knowledge of the law 5.19, 5.20
reasonable skill and care 2.29
responsibility for the work of others 5.81, 5.86
specialist expertise 2.24
estimates
  architects and engineers 5.23
  quantity surveyors 5.23
expert determination
  appointment of expert 8.61
  introduction 8.59
  liability to the parties 8.75
  obligations 8.65
expert witness
  Civil Procedure Rules, part 35 8.9
  meaning of 8.4
  duty to court 8.9, 8.19
  duties 4.22, 8.6, 8.9
  immunity from suit 8.23
  requirement for expert evidence in professional negligence claims 4.28
  appointment 8.21

fitness for purpose
  obligation in design 2.18

ICE *see* Institution of Civil Engineers
Institution of Civil Engineers 2.26
insurance
  aggregation 7.55
  claims made 7.9
  conditions 7.30
  contents of insurance policies 7.21, 7.34
  contract of 7.21
  deductible: *see* excess
  duty of fair presentation 7.36
  duty of utmost good faith 7.11, 7.36
  endorsements 7.23
  excess 7.22
  exclusions 7.5, 7.28
  insurance broker 7.46
  limit of cover 7.15
  notification 7.13, 7.47, 9.4
  policy of 7.21
  premiums 7.11, 7.21
  professional indemnity insurance 7.4
  proposal 7.34

QC clauses 7.61
refusal of indemnity 7.41,
schedule 7.22
third party rights 7.66

knowledge of the law
  architects and engineers 5.19, 5.20

legal proceedings 9.32, 9.59
liability
  contractual 3.1
  concurrency of tortious and contractual liability 3.77
  tortious 3.45
limitation
  operation of 6.75
limitation of liability
  architects and engineers 3.34
litigation 9.59
  County Court 9.61
  case management 9.62
  the claim 9.62
  costs of 9.66
  disclosure 9.62
  expert reports 9.62
  High Court 9.60
  offers to settle 9.67
  Pre-Action Protocol for Construction and Engineering Disputes 9.33
  pre-action steps 9.32
  statement of case 9.62
  Technology and Construction Court 9.61
  trial 9.62
  witness statements 9.62

mediation 9.9, 9.69

negligence *see* tort
net contribution clause 3.36

personal liability 3.128
professional
  meaning 1.3
professional bodies
  insurance requirements 7.1
  architects 2.7
  engineers 2.26
  project managers 2.53
  surveyors 2.40
project manager
  administration of contract 5.119
  Association of Project Management 2.53
  contracts 2.54
  function 2.50

meaning 2.46,
obligation to advise on contracts 5.116, 5.118
reasonable skill and care 2.47

quantity surveyor
  administration of contract 5.119
  advice on tender 2.36, 5.99
  contracts 2.41
  estimates 2.36, 5.23
  form of agreement 2.43
  function 2.32
  meaning 2.31
  reasonable skill and care 2.43
  Royal Institution of Chartered Surveyors 2.40,
  usual duties of 2.35
  valuations 2.30

reasonable skill and care
  architects and engineers 2.15
  codes of practice, relevance of 4.30
  contractual obligation 2.15, 2.30, 2.43, 3.31, 4.1
  expert determination 8.76
  expert witnesses 8.26
  express term of 2.15
  higher duty 2.16
  meaning of 4.1
  quantity surveyors 2.43
responsibility for the work of others
  architects and engineers 5.81, 5.86
RIBA *see* Royal Institute of British Architects
RICS *see* Royal Institution of Chartered Surveyors
Royal Institute of British Architects
  and architects 2.1, 2.7
  Standard Form of Agreement for the Appointment of an Architect 2.9
Royal Institution of Chartered Surveyors 2.40

standard of care *see* reasonable skill and care
sub-contract *see* responsibility for the work of others
surveyor *see* quantity surveyor

Technology and Construction Court 9.61
terms
  express terms of contract 3.16
  implied terms 3.40
third party
  duties owed to 3.55, 3.87, 3.91
tort
  architects' and engineers' duties to third parties 3.55, 3.87, 3.91

concurrency of contractual and tortious liability 3.77
duty of care *see* duty of care
negligence 3.45
tortious duties 3.45

utmost good faith 7.11, 7.36

warranty
architects' and engineers' warranties 2.17, 2.30
collateral warranties 3.108